Recommended Practice for Backflow Prevention and Cross-Connection Control

AWWA MANUAL M14

Third Edition

American Water Works Association

Science and Technology

AWWA unites the drinking water community by developing and distributing authoritative scientific and technological knowledge. Through its members, AWWA develops industry standards for products and processes that advance public health and safety. AWWA also provides quality improvement programs for water and wastewater utilities.

MANUAL OF WATER SUPPLY PRACTICES—M14, Third Edition

Recommended Practice for Backflow Prevention and Cross-Connection Control

Copyright © 1973, 1989, 2004 American Water Works Association

All rights reserved. No part of this publication may be reproduced or transmitted in any form or by any means, electronic or mechanical, including photocopy, recording, or any information or retrieval system, except in the form of brief excerpts or quotations for review purposes, without the written permission of the publisher.

Project Manager: Mary Kay Kozyra
Production Editor: Carol Stearns

Library of Congress Cataloging-in-Publication Data

Recommended practice for backflow prevention and cross-connection control.--3rd ed.
   p. cm. -- (AWWA manual ; M14)
  Includes bibliographical references and index.
  ISBN 1-58321-288-4
    1. Backsiphonage (Plumbing)--Prevention. 2. Cross-connections (Plumbing)--Standards.
  I. American Water Works Association. II. Series.

TD491.A49 no. M14 2004
[TH6523]
628.1 s--dc22
[696'.1]

                                                2003063633

Printed in the United States of America

American Water Works Association
6666 West Quincy Avenue
Denver, CO 80235-3098

ISBN 1-58321-288-4

Printed on recycled paper

Contents

List of Figures, v

List of Tables, vii

Acknowledgments, ix

Chapter 1   Introduction 1
    Purpose of Manual, 1
    Responsibilities, 1
    Health Aspects, 3
    Legal Aspects, 6

Chapter 2   Program Administration 11
    Types of Programs, 11
    Components of Water Purveyor's Program, 13

Chapter 3   Backflow Principles 29
    Basic Hydraulics, 29
    Types of Backflow, 31
    Assessing Degrees of Hazard, 34
    Assessing the Effectiveness of Assemblies and Devices, 38

Chapter 4   Backflow-Prevention Assembly Application, Installation, and Maintenance 41
    Backflow Control Methods, 42
    Backflow Devices, 56
    Testing, 59

Chapter 5   Typical Hazards 63
    Recommended Protection for Specific Categories of Customers, 63
    Recommended Protection for Water Purveyor's Hazards, 80

Appendix A   Assembly Test Procedures, 85

Appendix B   Model Ordinance, 109

Appendix C   Common Symbols, 117

Appendix D   AWWA Policy Statement on Cross-Connections, 119

Appendix E   Abbreviations of Agencies and Organizations, 121

Appendix F   Web Sites for Backflow-Prevention Incidents, 123

Glossary, 125

Index, 131

This page intentionally blank.

Figures

1-1   Examples of backflow-prevention equipment locations, 3
3-1   Schematic of a vacuum, 30
3-2   Backsiphonage backflow due to high rate of water withdrawal, 32
3-3   Backsiphonage backflow caused by reduced pressure on suction side of booster pump, 33
3-4   Backsiphonage backflow caused by shutdown of water system, 33
3-5   Backpressure backflow caused by carbon dioxide cylinder, 35
3-6   Backpressure backflow caused by pumping system, 36
4-1   Air gap on tank, 42
4-2   Air gap on lavatory, 43
4-3   Typical air-gap applications, 43
4-4   Additional typical air-gap applications, 44
4-5   Reduced-pressure principle backflow-prevention assembly, 45
4-6   Both check valves open and the differential relief valve closed, 46
4-7   Both check valves closed and the differential relief valve open, 46
4-8   Backpressure: both check valves closed and the differential relief valve closed, 46
4-9   Backsiphonage: both check valves closed and the differential relief valve open, 47
4-10 Typical reduced-pressure principle backflow-prevention applications, 48
4-11 Double check valve assembly, 49
4-12 Check valves open, permitting flow, 50
4-13 Negative supply pressure, check valves closed, 50
4-14 Backpressure: both check valves closed, 50
4-15 Typical double check valve assembly applications, 51
4-16 Pressure vacuum-breaker assembly, 53
4-17 Pressure vacuum-breaker assembly, normal flow condition, 53
4-18 Pressure vacuum-breaker assembly, backsiphonage condition, 54
4-19 Atmospheric vacuum breaker, 57
4-20 Dual check device, 58
5-1   Cross-connection control, water treatment plants, 83
5-2   Service-containment and area-isolation water treatment plants, 84
A-1   Major component parts of a five-valve differential pressure gauge, 88
A-2   Illustration of an RPBA test with a differential pressure gauge, 89

A-3	Illustration of a DCVA test with a differential pressure gauge, 91
A-4	Illustration of a PVBA test with a differential pressure gauge, 93
A-5	Shutoff with test adaptors for use on test cock, 94
A-6	Use of bypass hose, 95
A-7	Use of bypass hose for downstream shutoff test, 95
A-8	Illustration of pressure vacuum breaker, 96
A-9	Illustration of pressure vacuum breaker with differential pressure gauge, 97
A-10	Illustration of pressure vacuum breaker with differential pressure gauge (step 2), 98
A-11	Illustration of a double check valve assembly with differential pressure gauge, 99
A-12	Illustration of a double check valve assembly with differential pressure gauge (step 2), 100
A-13	Illustration of a double check valve assembly with differential pressure gauge—test for no. 2 shuoff valve test, 101
A-14	Illustration of reduced-pressure principle backflow preventer test (step 1), 103
A-15	Illustration of reduced-pressure principle backflow preventer test (step 2), 104
A-16	Illustration of reduced-pressure principle backflow preventer test (steps 3 and 4), 105
A-17	Illustration of SVBA, 106
A-18	Illustration of an SVBA test with a differential pressure gauge, 106

Tables

3-1   Means of backflow prevention, 40

5-1   Recommended protection at fixtures and equipment found in water treatment plants, 82

This page intentionally blank.

Acknowledgments

The importance of safeguarding our public water supply in the twenty-first century is more focused than ever before. This third (2004) edition of AWWA M14 brings the understanding of public health protection to a new level. Students of the second (1989) edition will observe significant changes to the text. Manual M14 now includes the practice of referring the reader to documents acceptable to the local jurisdiction, when appropriate to match local practice.

The AWWA Committee on Cross-Connection Control thanks recently retired committee chair Craig Adams, formerly of the Tucson, Arizona, Water Utility, for the leadership provided on both the Cross-Connection Control Committee and special ad hoc committee that developed the initial manuscript. Ad hoc committee members noted for their contributions include

Craig Adams
Mary Howell
Howard Hendrickson
Fred Baird
Patti Fauver

The AWWA Committee on Cross-Connection Control that reviewed and approved this manual had the following personnel at the time of approval:

Richard Coates, Chair, Miami–Dade Water & Sewer Department, Miami, Fla.

Craig Adams, City of Tucson Water, Tucson, Ariz.
Stuart Asay, Stuart Asay & Associates, Westminster, Colo.
Kenneth Ashlock, City of Tempe, Environment Division, Tempe, Ariz.
Fred Baird, Bac-flo Unlimited, San Antonio, Texas
George Bratton, Schaefer & Bratton Engineers, Coupeville, Wash.
Lou Allyn Byus, US Environmental Protection Agency, Jacksonville, Ill.
Pete Chapman, Wilkins Inc., Fresno, Calif.
Patti Fauver, West Jordan, Utah
William Foley Jr., Boston Water & Sewer Commission, Boston, Mass.
Harold Garrison, Kentucky American Water, Lexington, Ky.
John Halliwill, International Association of Plumbing & Mechanical Officials, Ontario, Calif.
Howard Hendrickson, Brentwood, N.H.
Mary Howell, Backflow Management, Inc., Portland, Ore.
Michael Kebles, Las Vegas Valley Water District, Las Vegas, Nev.
Kenneth Kelly, Severn Trent US, Inc., Colmar, Pa.
Mark Kneibel, City of Wyoming, Grand Rapids, Mich.
David McDonnel, Portland Water Bureau, Portland, Ore.
Charles Nena, California Water Service, San Pedro, Calif.
Les O'Brien, University of Florida TREEO Center, Gainesville, Fla.

Dan O'Lone, US Environmental Protection Agency, Atlanta, Ga.
Kanwal Oberoi, Charleston Commissioners of Public Works, Charleston, S.C.
Bruce Parrott, Watts Regulator Company, North Andover, Mass.
James Purzycki, BAVCO, Long Beach, Calif.
John Ralston, Louisville Water Company, Louisville, Ky.
Robert Rivard, MSHSIWAT, Lebanon, Conn.
Paul Schwartz, University of Southern California, Los Angeles, Calif.
Larry Stinnett, Knoxville Utilities Board, Knoxville, Ky.
Drennen Walker, City of Leeds Waterworks, Leeds, Ala.

AWWA MANUAL M14

Chapter 1

Introduction

PURPOSE OF MANUAL

This manual provides guidance to water purveyors on the recommended procedures and practices for operating a cross-connection control program. The purpose of any such program is to reduce the risk of contamination or pollution of the public water system.

A cross-connection is an actual or potential connection between any part of a potable water system and any other environment that contains other substances that, under any circumstances, would allow such substances to enter the potable water system. Other substances include gases, liquids, or solids, such as chemicals, water products, steam, water from other sources (potable or nonpotable), and any matter that may change the color or taste of water or add odor to water.

RESPONSIBILITIES

In the United States, the federal government, under the Safe Drinking Water Act (SDWA) 42 U.S.C. § 300f to 300j-26, has jurisdiction over the public health aspects of the drinking water supply. State governments also have jurisdiction over matters of public health related to the supply of water. The state regulations cannot supersede the federal regulations; however, they may be more stringent than the federal regulations. Lower levels of government within a state, with the authority of the state, may impose other regulations or more stringent regulations not in conflict with state regulations.

In Canada, the provincial governments have jurisdiction over the public health aspects of the drinking water supply. Lower levels of government within a province (e.g., regional districts and municipalities), with the authority of the province, may impose other regulations or more stringent regulations not in conflict with provincial regulations.

Because there is a difference between the authority of the United States federal government and that of Canada and between the different states and provinces, the following discussion, although referring to "federal and state," illustrates the different regulations governing a public water utility.

The federal and state governments regulate the public health aspects of the supply of drinking water "to protect the health, safety, or welfare of the users of water." These regulatory or "police" powers govern the manufacturers and distributors of water (e.g., water utilities, bottled water plants, etc.). Regulations apply to the entity that owns (controls) the facilities for the manufacture and/or distribution of water.

For US water utilities (water purveyors), the SDWA regulations govern public water systems. The SDWA (see 42 U.S.C. 300f(4)(A)) states: "The term *public water system* [PWS] means a system for the provision to the public of water for human consumption through pipes or other constructed conveyances, if such system has at least 15 service connections or regularly serves at least 25 individuals. The public water system includes (i) any collection, treatment, storage and distribution facilities under control of the operator of such system and used primarily in connection with such system, ..."

A government body that owns a water utility (e.g., a city) is regulated by the higher levels of government. All water utilities manufacture or provide a product (potable water) and, like other industries, are regulated by government. Except for certain types of noncommunity systems where water consumption is solely within the property owned by the water purveyor, the federal, state, and local regulatory jurisdiction over a public water system ends at the water purveyor's point of service. Downstream of the point of service, federal, state, and local responsibilities "to protect the health, safety, or welfare of the users of water" fall under the jurisdiction of agencies other than those regulating public water systems, and include, but are not limited to, the following:

- Building and plumbing officials who are responsible for regulating the potable water system (plumbing) past the point of service. Their work generally consists of reviewing plans and inspecting plumbing of permitted new or remodeled construction.

- Fire marshals who are responsible for regulating fire protection systems (e.g., fire sprinkler systems) downstream of the potable water system supply connection entering the premises.

- Safety inspectors (Occupational Safety and Health Administration [OSHA], Workers' Compensation Board [WCB] [Canada], Mine Safety and Health Administrators [MSHA]) who are responsible for inspecting potable water systems (plumbing) for workers' safety.

- Health officials who are responsible for inspecting restaurants and other food preparation facilities (e.g., dairies), health care facilities (e.g., nursing homes), etc.

- Agricultural inspectors who are responsible for the safe handling of chemicals (e.g., pesticides) used in growing and processing agricultural products.

These agencies have jurisdiction over all work on the customer's premises. All have regulations that involve cross-connection control. These different regulations may be in conflict with the procedures for cross-connection control. The authority of these agencies over the water purveyor's customers may be continuing or may be limited by the issuance of a final permit (e.g., for building occupancy) (see Figure 1-1).

The water purveyor is not given authority to exercise any of the above-noted regulatory responsibilities without an agreement for an assignment of authority. Thus, the water purveyor has no authority, hence no responsibility, beyond the meter for the customer's compliance with cross-connection regulations.

The water purveyor's responsibility is to supply potable water to its customers at the point of delivery (e.g., water meter). The characteristics of potable water are defined by regulations. The testing of water quality (e.g., for lead and copper levels) may involve collecting water samples on the customer's premises. However, this does not impose a responsibility on the water purveyor for regulating plumbing. As stated in the USEPA Safe Drinking Water Act, "Maximum contaminant level means the

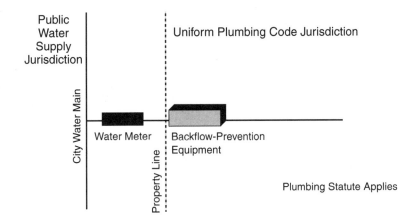

A. Location on private side of property line

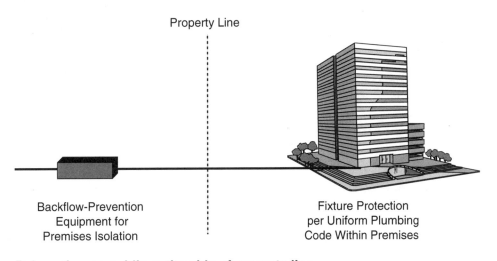

B. Location on public entity side of property line

Courtesy of Washington Department of Health

Figure 1-1 Examples of backflow-prevention equipment locations

maximum permissible level of a contaminant in water which is delivered to any user of a public water system."

HEALTH ASPECTS

Protection of drinking water for public health emphasizes preventing contamination. A multiple-barrier approach is used from the source to the tap. The following are major barriers established for public water systems:

- **Sources of supply:** Prevent human contaminants such as viruses and bacteria or chemicals from entering the water supply through watershed control and wellhead protection programs.
- **Treatment:** Remove or reduce natural and human contaminants to comply with the maximum contaminant levels (MCLs) established by regulations.

- **Chlorination:** Maintain chlorine residual in the water supply to control microbiological quality.
- **Storage:** Provide covered storage and prevent microbiological contamination through openings in reservoirs.
- **Distribution:** Comply with installation and material standards and provide minimum operating pressures to prevent contaminants from entering the system.
- **Cross-connection control:** Provide premises isolation (containment of service) or equivalent in-premises fixture protection to prevent contaminants from entering the water purveyor's system.
- **Water quality monitoring:** Provide surveillance of system to detect contaminants in the water supply.
- **System operator:** Ensure that qualified personnel operate public water systems through operator certification.
- **Emergency plan:** Establish procedures for correcting problems detected in water quality monitoring or caused by natural disasters.

On the customer's premises, plumbing and health codes establish minimum design, installation, and operating requirements for public health protection. Major items in the plumbing codes are as follows:

- **Distribution:** Install approved materials and follow design requirements to ensure adequate pressure at fixtures.
- **Cross-connection control:** Provide backflow preventers at fixtures to prevent contaminants from entering the system.
- **Licensed plumber:** Require that a licensed plumber (with some exceptions) install and repair all plumbing.

These requirements are conservative. They include a high safety factor for system design (reliability) and for acceptable contaminant levels. For example, regulation of chemical contaminants may be based on a possible adverse health effect from the long-term (e.g., lifetime) consumption of 2 liters of water per day with a chemical at a level above the MCL.

Most SDWA requirements deal with possible chronic (long-term) health effects. Contamination of a water distribution system through a cross-connection often results in acute (immediate adverse) health effects that may result in illness or death of one or more persons and/or financial losses. Although cross-connection control is only one of the multiple barriers to protect water quality, it is one of the most important. Without the water purveyor's cross-connection control program, the distribution system may become the weak link in the multiple-barrier approach.

Potable water is water that does not contain objectionable pollution, contamination, minerals, or infectious agents and is considered satisfactory for drinking. By this definition, potable water need not be pure or ultrapure. Potable water may contain bacteria (e.g., low levels of heterotrophic bacteria) and other contaminants. For cross-connection control purposes, potable water is considered to be safe for human consumption, i.e., free from harmful or objectionable materials as described by the health authority. In assessing the degree of hazard, "safe for human consumption" or "free from harmful or objectionable materials" are not clearly defined parameters. A chemical toxin in high concentrations may cause no harm when consumed in low concentrations.

In assessing the actual and potential degrees of hazard, microbiological, chemical, and physical parameters must be considered. These parameters are described in the following paragraphs.

Microbiological

Waterborne diseases are the primary concern in cross-connection control. Waterborne diseases are caused by the following major groups: bacteria, virus, algae, fungi, protozoa, and parasitic helminths (worms). The risk to public health of a waterborne disease transmitted through the public water supply is exacerbated by the

- large population that may be exposed to the disease;
- inability to immediately detect contamination (The first indication may be the outbreak of disease.); and
- difficulty in tracing the source (For example, *Giardia* cysts may enter the distribution system from a reservoir or through a cross-connection with an auxiliary supply.).

Contributing to the difficulty of assessing the relative risk to public health from a microbiological contaminant is the issue of infectious dose. The health effect to an individual consuming a microbiological contaminant varies by the type of organism, the quantity ingested, and the strength of the person's immune system. For example, water with a low level of the total coliform bacteria *Citrobacter freidii* presents little adverse health concern; however, this bacteria often colonizes distribution system piping. By comparison, the ingestion of only a few *Giardia* cysts may be infectious.

Although a microbiological contaminant may not be a pathogen or opportunistic pathogen (one that affects a person with a weak immune system), their presence in the water distribution system may be an indirect concern. Some microbiological contaminants may cause taste and odor problems or cause a chlorine demand. Any coliform bacteria detected in the water purveyor's monitoring program will require mitigation measures that vary from resampling to a boil-water order with an emergency water main flushing and disinfection program.

In assessing the problem of bacteria entering the distribution system, the water purveyor must consider the following issues:

- Poor-quality source water may enhance bacteria growth and regrowth in the distribution system. For example, source water with a high level of organics provides a food source for bacteria that may enter the distribution system due to a backflow incident. Other quality concerns include water with high turbidity, sulfate-reducing bacteria, and iron and manganese that provide a biofilm (slime) or biomass (sediment) in water mains that facilitate bacteria regrowth.
- Distribution system piping that is in poor condition may aid bacteria growth and regrowth. For example, corrosive water may cause tuberculation to form on old unlined cast-iron and steel water mains. The tubercles provide a rough surface that shelters bacteria from chlorine.

    Systems may have inadequate capacity to maintain pressure during peak water demand periods (e.g., fire flow, hot summer weather). Many old distribution systems have a relatively high frequency of breaks or leaks. Whenever there is a reduction or loss of pressure in the distribution system, there is the likelihood that contaminants will flow back into the potable water system.
- The inability and unwillingness to maintain a chlorine residual in the distribution system makes it possible for bacteria to live and grow.

Because each water system is different, the concerns about microbiological contamination are different for each water purveyor.

Chemical

Every chemical has an effect on the living organisms that are exposed to it. The higher the dose, the more significant the effect; and the longer the exposure, the more significant the effect. Consequently, the toxic dose of each chemical must be considered. Acute toxic conditions are of most concern in cross-connection control, because immediate health impacts may result.

The health effects of a toxic chemical vary by type of chemical, quantity ingested, and the immune system of the infected person. For most people, ingestion of water with a high copper level will likely cause nausea, diarrhea, abdominal pain, and/or headache. In the small portion of the population that is allergic to copper, the health effects are far worse, perhaps causing death.

Some chemicals have a low level of toxicity. However, when combined with the chemicals that are added to a water supply, a more toxic chemical may form. Chemical contaminants may also react with the piping material in the plumbing or distribution system to leach toxic metals into the water. Because every water system treats its water differently, concerns about chemical contamination are different.

Physical

There are few physical hazards that are not also chemical hazards. Examples of "pure" physical hazards include hot water and steam. Human contact with these hazards may result in burning of the skin, eyes, etc. However, the public health risk of a physical hazard, such as propane gas, offers a different degree of hazard. In addition to their toxic effects, chemicals may cause hazards such as an explosion (propane gas) or damage to piping material (e.g., permeation of polyvinyl chloride [PVC] pipe, leading to a weakening of the pipe and then to structural failure).

LEGAL ASPECTS

The following discussion provides general information for the utility manager. Each utility should consult a qualified attorney prior to establishing risk and liability management policies and procedures related to their cross-connection program. The following information includes some excerpts from the AWWA–Pacific Northwest Section manual *Cross-connection Control Program Administration*.

Eliminating all cross-connections is an enormous task, one that could require resources beyond the financial capacity of many water systems, as well as public health and plumbing inspection departments. Once contamination from a cross-connection occurs, it is likely that one or more persons will suffer some type of loss, e.g., a minor financial loss to cover the cost of flushing a plumbing system or serious injury or death and resulting social and economic damages.

Given the magnitude of potential liability in this area, three important areas of legal impact that are likely to influence a utility's legal coverage include (1) contractual responsibilities; (2) governmental statutes, regulations, and local controls; and (3) common law doctrines, which form the basis for civil liability to third parties.

Contractual Responsibilities

In general, when an entity assumes the role of water purveyor and a property owner accepts service from this entity, an implied contract is formed. In some instances, this contractual relationship is formalized by having a customer sign an application for service. Once a contract is formed, obligations must exist between both parties—the water purveyor's obligation is to provide uninterrupted service (except for necessary repairs, maintenance, or nonpayment) and potable water, and the customer's obligation is to pay for the service and comply with reasonable rules and requirements established by the water purveyor. To ensure a customer's obligation to comply with these rules and requirements, the rules should be in writing and accepted in writing by the customer; the water purveyor should send out updates to these rules at least annually. Qualified counsel should be consulted concerning legal requirements for enforceability of these rules.

Note that these rules also bind the water purveyor as part of the water service agreement. As part of the agreement to supply water, the water purveyor may stipulate conditions for maintaining water supply to the customer. This may include the customer's agreement to allow the water purveyor to inspect the premises for cross-connections and the customer's agreement to take prescribed measures to protect the water system from contamination. A customer's failure to honor the contract may be grounds for the water purveyor to discontinue water supply to that customer.

The water purveyor must specify reasonable conditions for the supply of water. With regard to backflow prevention, the water purveyor normally establishes conditions for service ranging from a list of approved backflow-prevention assemblies to preapproval of backflow-assembly testers. These conditions for service may impact the customer's cost of obtaining water service, the profit for suppliers of goods relating to cross-connection control (e.g., manufacturer of backflow assemblies), and the income for suppliers of services relating to cross-connection control (e.g., backflow-assembly testers). A person or entity suffering a loss due to a water purveyor's unreasonable conditions for service may be permitted to seek to recover damages in a court of law or before a state public utility commission.

In establishing conditions for service, the water purveyor should be able to show the following:

- It has a legitimate interest in establishing the requirements. Although a water purveyor may wish to protect its customers from harming themselves, this is the responsibility of regulatory authorities. The water purveyor's primary interest is to protect its distribution system from contamination.
- The conditions for service are not arbitrary, i.e., they can be defended as reasonable based on the standards of the water industry.
- The conditions for service are not exclusive with regard to the supply of goods or services.

Additional information on establishing reasonable conditions for service is provided in chapter 2.

A homeowner or customer may contract with a third party, e.g., a contractor or builder, to perform certain obligations under a service contract with the water purveyor (i.e., comply with cross-connection program regulations). However, this does not relieve the homeowner or customer of his or her responsibilities for legal compliance. Some contractors use agreements that attempt to limit their obligations or liabilities to the homeowner or customer. However, these agreements generally do not limit the customer's obligations to the water purveyor. A water purveyor may work with a customer's selected contractor, agent, or employee to explain program requirements, but the customer must always be advised that he or she remains responsible to the water purveyor.

In addition to the terms of any contract, the water purveyor may be required by state, county, or city regulations to discontinue service to a customer that fails to comply with government cross-connection control regulations. The legal aspects of such statutory and regulatory requirements are addressed briefly in the following section.

Government Statutes, Regulations, and Local Controls

Federal and state legislative bodies are heavily involved in adopting statutes that have a major impact on drinking water purveyors. Appropriate administrative agencies also promulgate regulations and periodic regulatory changes pursuant to their statutory authority. Local governments also may impose controls over water purveyors through ordinances, regulations, rules, orders, permits, licensing, and the like.

Upon proper enactment of statutes and local ordinances and promulgation of regulations and other administrative actions, regulated entities are deemed responsible for knowing and obeying these laws. Although most government agencies make an effort to notify affected parties of their newly established and ongoing obligations, contractors and builder groups should be involved in the process of enacting new laws and developing regulations for their implementation.

The primary federal statute governing the safety of public water systems in the United States is the SDWA. Although major portions of the SDWA, as amended, deal with information gathering, source protection, and PWS assistance, federal and state enforcement of National Primary Drinking Water Regulations (NPDWRs) is also authorized. Noncompliance can be the subject of an administrative order or civil action, both potentially involving significant civil penalties of up to $25,000 per day of violation. Criminal sanctions for certain "willful" violations of SDWA requirements are also possible. Although backflow incidents that result in the delivery of contaminated water are to be distinguished from a PWS's failure to routinely meet established MCLs, these same criminal statutory sanctions may be applied to such an event.

The SDWA's reporting requirements may also apply to a backflow incident, whether it is the subject of enforcement or not. A variety of circumstances and events, such as failure to comply with an MCL, must now be reported to those served by a PWS (see 42 U.S.C. 300g-3(c)). This type of required disclosure is a strong deterrent, even in the absence of civil penalties, because it exposes a water purveyor to a third-party lawsuit under other statutory and common law.

A variety of state statutes that apply to water purveyors could be triggered by cross contamination due to a backflow incident. The following is an example of a statutory requirement from the state of Washington, Section RCW 70.54.020, Furnishing Impure Water—Penalty:

> Every owner, agent, manager, operator or other person having charge of any waterworks furnishing water for public use, who shall knowingly permit any act or omit any duty or precaution by reason whereof the purity or healthfulness of the water supplied shall become impaired, shall be guilty of a gross misdemeanor. [1909 c 249 {291;RRS} 2543]

Other laws and regulations that impact the water purveyor include

- federal and state environmental and consumer protection regulations, including product liability laws (e.g., supply of tainted product: contaminated water); and
- state requirements for the implementation of a cross-connection control program, testing of assemblies by certified testers, reporting of backflow incidents, records, etc.

Water purveyors should remain aware of applicable state and local laws and regulations and consult qualified legal counsel concerning their possible application in the case of a backflow incident.

Common Law Doctrines

A common law duty of every water purveyor is to supply potable water to its customers. A water purveyor's cross-connection control program should be designed to reasonably reduce the risk of contamination of the utility's system and the water purveyor's exposure to legal liability. If it is determined that a water purveyor has failed to meet this duty, the water purveyor could be held liable to its customers for damages proximately caused by the water purveyor's breach of this duty. If other parties (contractors or other individuals) are at fault, their liability to any injured party may be determined in a similar manner, with any party found to have caused damage to another assessed damages for some and possibly all injuries suffered.

Liability for supplying impure water has long been recognized as common law, most often for the incident of disease or poisoning that results from the violation, as well as for damage to machinery and goods suffered by commercial customers. Although case law varies from state to state, the general standard created by these cases is one of exercising reasonable or ordinary care to furnish pure water. Liability may result if this duty to exercise reasonable care and diligence in supplying water is breached. However, some other cases emphasize that water suppliers are not insurers or guarantors of the quality of water supplied by them.

With respect to damages, a customer has the burden of proving all "special" or specific damages, such as reasonable and necessary medical expenses that were incurred as a consequence of the asserted breach of duty. Property damage, lost market value to property, loss of income, court costs, and any other specific items of expense may also be claimed. In addition, a customer may seek an award of general damages, which are for general pain, suffering, and discomfort, both physical and emotional. There is no precise formula for computing these kinds of damages, and unless a law is in place to limit damages, they are determined by a judge or jury after considering the evidence introduced at trial.

If statutory performance standards such as those included in the SDWA are violated in connection with a backflow incident, the water purveyor's noncompliance with such standards will ease a claimant's burden of proof. This, in turn, will allow a claim of negligence per se for having violated the standard but will still require a showing that a customer's claimed damages were proximately caused by the instance of noncompliance. Because the SDWA's record-keeping and reporting requirements generate a large quantity of data on which such actions might be based, it is more important than ever to implement backflow-prevention measures, rather than simply defending against government enforcement or third-party claims of damage.

In summary, a backflow-prevention program should

- comply with regulatory requirements and
- follow these "standards of the water industry" for the application of backflow assemblies:
  — the installation of approved backflow-prevention assemblies, with the approval of assemblies being based on conservative industry standards;
  — the installation of backflow methods, assemblies, and devices in accordance with industry standards;
  — the testing of backflow assemblies by a procedure and at a frequency stipulated in regulations and recommended by industry standards;
  — the testing of backflow assemblies by certified testers, where the certification establishes a minimum level of training of the personnel testing assemblies;
  — isolation of premises (containment of service) by the installation of a backflow assembly on the customer's water service;
  — procedures for investigating backflow incidents and restoring water quality;
  — entering into a written contract with the customer to provide water service; and
  — a comprehensive record-keeping program.

This page intentionally blank.

AWWA MANUAL M14

Chapter 2

Program Administration

For millennia, people have been concerned with obtaining and maintaining pure and wholesome water supplies. Archeological studies reveal that as early as 3000 B.C., the ancient Egyptian State had a government official who was required to inspect the country's water supply every 10 days. With the widespread use of water closets in the 1800s came direct cross-connections with water mains. This brought into focus the problem that, as one nineteenth century authority stated, "foul matters may get into the pipes."* Today, all areas of government and industry are aware of the need to prevent contamination of potable water supplies through cross-connections. However, the goals and levels of involvement vary.

TYPES OF PROGRAMS

Cross-connection control programs may be undertaken by, but are not limited to, the following:
- Property owner (e.g., plant safety committee program)
- Building and plumbing inspectors
- Health inspectors
- Safety inspectors (e.g., workers' compensation board)
- Water purveyors

Their cross-connection control efforts may include
- A cursory inspection to determine compliance with regulations. The owner of a potable water system must comply with regulations notwithstanding the failure of the inspector to find a code violation.

---

*A.J. Keenan, C.S.I., B.C. Section AWWA Cross-connection Control, September 1977.

- A detailed inspection to ensure compliance with construction specifications, provide workers with protection, etc.
- A survey to assess the overall public health risk or identify major cross-connection hazards
- A survey to locate the source of a detected contaminant

Cross-connection control may be divided into two categories:

- Prevention of contamination of the facilities under the owner's control
- Regulation of the owners of potable water systems

The water purveyor and the water purveyor's customers fall under the first category. The goal of each is to reduce the risk of contamination and potential liability from backflow into the potable water system they own.

The water purveyor may protect its system by

- Isolating its customer through the installation of a backflow preventer on the service pipe. This procedure is known as "service protection," "containment," or "premises isolation." The backflow preventer may be
    — owned by the water purveyor: most frequently installed upstream of the point of service, or
    — owned by the customer: required by the water purveyor to be installed downstream of the point of service.
- Relying on the installation of backflow preventers at each plumbing fixture in the customer's system. This is referred to as "internal protection," "fixture protection," or "in-premise protection."

Where the water purveyor adopts a service-protection type of cross-connection control program, each customer must be evaluated to determine the overall health risk to the public water system imposed by the customer's plumbing system. The evaluation may involve a survey of each customer's premises.

Where the water purveyor adopts internal protection and relies on the customer's internal fixture protection, the plumbing should be surveyed to determine if the protection provided is satisfactory to the water purveyor. The water purveyor may rely on compliance with the plumbing code when assessing the overall risk to the public water system. The water purveyor may also impose additional requirements for internal fixture protection as a condition of providing service without the installation of a backflow preventer on the service. In either case, the water purveyor's customer must understand that the water purveyor is not "inspecting" the plumbing to determine compliance with regulations or to ensure protection of the occupants of the premises.

If the water purveyor imposes additional requirements for internal protection (fixture protection), the requirements must be reasonable with respect to what is needed to protect the water purveyor's system. For example, it would not be reasonable for the water purveyor to require all plumbing fixtures to be isolated with a backflow-prevention assembly that provides the highest level of protection in lieu of placing one backflow-prevention assembly on the water service.

Regardless of the type of program selected, it must be understood that the water purveyor is responsible for any contaminant that enters its distribution system. Reliance on the customer to prevent contamination at the fixture increases the water purveyor's risk and potential liability.

Property owners who are also the water purveyor of a property generally implement an internal protection program because they are liable for both the distribution and private plumbing systems. Because internal protection is typically regulated by plumbing codes, this manual focuses on water purveyors that supply water to customer-owned properties.

COMPONENTS OF WATER PURVEYOR'S PROGRAM

The recommended components of a water purveyor's cross-connection control program are discussed in this section. For convenience, the components have been grouped into several categories based on administrative responsibility, service policy, and related program components.

Establishing Authority and Administrative Responsibility

The first step in implementing a cross-connection control program is for the water purveyor's management group to formally establish the program. Program operation involves making a commitment to employ staff, establishing policies that involve risk and liability decisions, setting program goals, etc. A program may be established by resolution or ordinance, as outlined in the sample ordinance in appendix B. The program may also be established by incorporating it into an adopted water system plan.

One person should be selected as program administrator. The various tasks in operating the program may be performed by different staff members, consultants, etc. However, to ensure consistency and accountability, one person must be in charge.

Management should periodically review the program's performance. It may be necessary to make changes in program administration and policies if the program goals are not being met or if the program needs to be expanded.

Establishing the Service Policy

When establishing service policies, the water purveyor should consider service protection, containment, and premises isolation. The water purveyor must decide on the extent to which it will rely on the customer's internal fixture protection. This decision will be governed in part by state or provincial regulations that establish mandatory service protection.

Mandatory service protection is usually required for high-hazard categories of customers, such as

- radioactive material processing plants or nuclear reactors;
- sewer treatment plants, sewage pump stations, or waste dump stations;
- hospitals; medical centers; medical, dental, and veterinary clinics; and plasma centers;
- mortuaries;
- laboratories;
- metal-plating facilities;
- food-processing and beverage-bottling facilities;
- car washes;
- premises with an auxiliary water supply;
- premises where access is restricted;
- piers and docks, graving docks, boat marinas, dry docks, and pump stations;
- premises with fire sprinkler systems and/or private fire hydrants; and
- irrigation systems.

The water purveyor must ensure that a backflow preventer acceptable to the state or province is installed on the potable water service to these premises. The recommended type of backflow preventer for service protection and the reasons for the risk assessment are outlined in chapter 5.

If the state or provincial authority has not established a list of water users for mandatory service protection, the water purveyor should do so. The water purveyor may augment any list established by the state or province, unless specifically prohibited by regulations. In augmenting the list for service protection, the water purveyor may specify the type of backflow preventer. The options available to the water purveyor range from requiring a backflow preventer on all service connections to requiring a backflow preventer only on the service to a category of water user or customer listed by the state or province for mandatory service protection.

Where a service protection backflow preventer is not installed, the water purveyor has the responsibility to periodically reassess the risk. Plumbing may be changed without permit and a hazard created. This responsibility to periodically reassess a customer also applies if the maximum backflow protection is not provided for service protection. The backflow preventer must be commensurate with the assessed degree of hazard. It must be in a safe location, away from toxic fumes, and should be protected from the elements. If the maximum protection is not provided, a change in the customer's plumbing may require a change in the type of backflow preventer relied on for service protection.

Where the water purveyor relies on the customer's internal fixture protection, the effort to assess and reassess the risk to the water purveyor is much more complex, and thus, time consuming. This is an important consideration when establishing a policy to rely on internal fixture protection.

The main benefits of accepting internal fixture protection in lieu of containment are (1) the customer may benefit by saving the cost of a service protection backflow preventer, and (2) the water purveyor's risk assessment survey may assist the customer by ensuring adequate internal fixture protection.

The main disadvantages to the water purveyor of relying on the customer's internal fixture protection are an increased risk of contamination of the distribution system (e.g., the failure to identify a cross-connection hazard, changes made to plumbing unknown to the water purveyor) and an increase in program cost (e.g., for surveys to reassess the customer, monitor numerous backflow preventers rather than one on the service).

Note that the customer is the primary beneficiary of the water purveyor's reliance on fixture protection, while the water purveyor potentially accepts all the risk and liability therefrom.

Ownership of Service Protection Backflow-Prevention Assemblies

The water purveyor may install the service protection backflow preventer along with its water meter. The main advantages to having the water purveyor own the backflow preventer are

- assurance that the backflow preventer will be properly field-tested and maintained;
- assurance that the backflow preventer is appropriate for the application;
- the cost of installation, testing, and maintenance may take advantage of the economies of scale (e.g., purchase of a large number of units should provide a volume discount);
- the cost of the backflow preventer may allow for an increase in rates (for investor-owned utilities); and
- less need to enforce the water purveyor's policies for the customer's failure to field-test and maintain the backflow preventer.

The main advantages to the water purveyor of the customer installing the backflow preventer are

- all costs will be directly borne by the customer;
- the customer will have the responsibility, and the liability, for the proper installation, field testing, and maintenance of the backflow preventer;
- the backflow preventer may be installed in the customer's building where protection is provided from freezing and vandalism; and
- the customer can select the backflow-prevention assembly manufacturer that is best for their water service conditions (e.g., pressure loss, size, orientations, etc.).

Testing Backflow-Prevention Assemblies

To ensure that a backflow preventer functions properly, it should be field-tested after installation and periodically thereafter, as discussed later in this chapter. The water purveyor is responsible for ensuring that the backflow preventers relied on to protect its distribution system are field-tested. The types of backflow preventers designed for field testing, referred to as *backflow-prevention assemblies*, are discussed in chapter 4. Backflow-prevention assemblies owned by the water purveyor may be field-tested by staff or by a contractor employed by the water purveyor.

The main advantages of having the water purveyor field-test backflow-prevention assemblies are

- improved quality assurance,
- provision of service to the customer, and
- reduced administrative task of issuing notices to the customer to field-test backflow preventers.

The main advantages to the water purveyor of having the customer field-test backflow-prevention assemblies are

- the cost of field testing is borne by the customer,
- field testing may be coordinated with maintenance or other work,
- the customer has the responsibility and thus the liability for the field testing, and
- the customer is aware of the importance of backflow prevention.

Enforcement Action

Historically, the primary enforcement action used by water purveyors has been to shut off service. Although immediately effective in alleviating the risk of backflow, the termination of water service can create a health hazard and substantial financial loss for the customer. Water is needed for sanitation (e.g., flushing of toilets), and the shutoff of a service may create a backflow condition within the customer's plumbing. Denial of water service to a commercial customer may result in a substantial loss of business.

Although the water purveyor may have the right and does have the obligation to shut off service to prevent backflow, the exercise of this right must be reasonable. A customer's failure to submit a field-test report within 30 days does not constitute an immediate risk to the water purveyor. Any enforcement action by the water purveyor made on the grounds of a "public health risk" may be challenged by the customer in a court of law. If the actions of the water purveyor are found to be unreasonable, the water purveyor may have to pay substantial compensatory and punitive damages.

For these reasons, the water purveyor should establish clear guidelines for enforcement actions, which may include installation by the water purveyor of a service protection backflow preventer on the service pipe rather than discontinuing water service. The enforcement policy should address the following:

- State and local laws and/or regulations;
- The effort to be made to notify the customer of the requirement before the enforcement action is taken (For example, mail first notice, mail second notice 30 days thereafter, mail final notice 15 days after the second notice. One [or more] notice[s] should be sent via certified mail, return receipt requested.);
- The appeal procedures of any enforcement action;
- The criteria for immediate shutoff of water (e.g., meter running backward, backflow occurring);
- Notification of customer noncompliance to other authorities (e.g., state/provincial health department); and
- Variation in policies for different water services (e.g., fire lines) or class of customers (e.g., apartment buildings).

The water purveyor should consult an attorney when establishing an enforcement policy and prior to taking any enforcement action that could cause a major financial loss or operating problem for a customer.

When establishing an enforcement policy, the water purveyor should consider the advantages of a written service agreement with the customer. The customer's failure to comply with the water purveyor's requirements for the installation and maintenance of backflow preventers would constitute a breach of contract by the customer. Remedial action (e.g., shutoff of service) would be specified in the contract.

Water Purveyor's Risk Assessment

Before water service is provided to a new customer, the water purveyor should perform a hazard assessment and establish the requirements for backflow prevention. The methods of assessing risk and selecting a backflow preventer commensurate with the degree of hazard are discussed in chapter 3.

Although new water services can be easily surveyed for degree of hazard, this does not address customers who had water service before the cross-connection control program was implemented. To ensure that all hazards are identified and properly protected, a program to survey existing water users is recommended. This program includes

- assessing the risk from existing customers,
- notifying customers of the assessed risk and required backflow preventer(s),
- ensuring that backflow preventers are installed on existing services or at internal fixtures in lieu of service protection, and
- periodically reassessing the risk.

The water purveyor must establish a schedule for accomplishing these. Priority must be assigned based on degree of hazard. For risk and liability management, the water purveyor should give priority as follows:

1. Customers requiring mandatory service protection
2. Commercial customers, industrial customers, fire protection customers
3. Multifamily residential customers
4. Single-family residential customers

The priority list may be further refined with subcategories as needed for scheduling work.

Management should establish the schedule for completing the above tasks. Until the cross-connection control program is fully developed, the water purveyor is at risk. The decision to take a risk and incur potential liability should only be made by management.

For public relations reasons and for risk and liability management, the water purveyor should consider first providing backflow protection at all of the facilities it owns (e.g., water treatment plant, works yard). The recommended protection for these facilities is discussed in chapter 5.

Testing of backflow-prevention assemblies. Most backflow preventers designed for field testing can be tested in place. The water purveyor should field-test or require to be field-tested the backflow-prevention assemblies it relies on under these circumstances:

- on installation,
- at least annually,
- after repairs,
- after relocation or replacement, and
- on responding to a reported backflow incident.

The water purveyor must specify the field-test procedures it will accept. As a minimum, state or provincial procedures must be followed. Although a uniform procedure within a state or province is desirable, the water purveyor may include supplemental field tests to maintain operability and assure performance. Because backflow preventers must be installed in a manner that will facilitate field testing, the water purveyor should either adopt installation requirements as part of its service policies or make reference to industry or association standards and specifications that incorporate installation requirements.

Training

Proper field testing of backflow preventers and assessment of hazards requires specialized training and/or experience. Backflow-prevention assembly testers and cross-connection control program administrators can be trained at various locations throughout North America. Training includes seminars, workshops, conferences, and courses and may be provided by states or provinces, technical organizations (e.g., AWWA sections), colleges, manufacturers, consultants, and individuals.

Training is available for

- cross-connection control program administration,
- backflow-prevention assembly field testing,
- backflow-prevention assembly repair, and
- cross-connection surveys and inspection.

Not all training courses are equal. Training courses within North America, a geographical region, or even a state or province may vary in the technical content, instructor's qualifications, course length, etc. For example, training courses may vary in scope, ranging from an introduction to cross-connection control to basic training to continuing education to advanced training. (Benchmark training will, at a minimum, total approximately 32 hours.) Some training courses may be directed to specific groups (e.g., plumbing inspectors).

Some states or provinces may review the training to establish adequate quality and ensure that it is consistent with their cross-connection control regulations. Other states or provinces rely on technical organizations (e.g., AWWA sections), universities, manufacturers, consultants, and individuals to provide training.

In selecting training for their staff, the water purveyor should consider the following:

- Where state or provincial courses are available, determine whether the content is satisfactory and whether supplemental training is appropriate to improve quality assurance.
- Where state or provincial courses are not available, evaluate and select the other sources of training, taking into consideration such things as the official recognition of the training course by the state or province; the scope and content of the subject matter; the instructors' qualifications; and the quality of the training materials, facilities, and equipment.
- Provide supplemental continuing education through seminars, workshops, refresher courses, and conferences on cross-connection control.
- Provide publications relating to cross-connection control as an important source of supplemental continuing education (see appendix E).
- Consider participating in regional cross-connection control technical groups to exchange current information (e.g., an AWWA section's cross-connection control committee).

Certification

Cross-connection testers should be certified to ensure the minimum level of proficiency needed to perform the task of testing backflow-prevention assemblies. A certification program may be administered by a state or province, a technical organization (voluntary program), local administrative authority, or the water purveyor. Where the state or province requires certification, the water purveyor may require additional certification to ensure a higher level of proficiency.

As a minimum, the water purveyor should require that certification for backflow-prevention assembly testers (staff and/or contractors submitting field-test reports) includes

- successful completion of a certification exam acceptable to the administrative authority (Such exams incorporate both written questions and hands-on testing of backflow-prevention assemblies overseen by a competent proctor independent from the training course.),
- periodic recertification through a written and hands-on exam, and
- procedures for suspending or revoking certification for tester misconduct or improper testing of assemblies.

Legal considerations of training and certification. Both training and certification have legal implications that warrant consultation with a qualified attorney. Water purveyor staffs should avoid personal defamation and/or product disparagement with reference to persons or entities providing training and certification. Conduct constituting the unlawful restraint of trade or other violations of federal and/or state antitrust laws should be carefully avoided.

Training courses and certification should always be selected based on reasonable criteria and should not be arbitrary or discriminatory. The following serve as examples:

- The water purveyor should consider whether certification or other proof of qualifications required as a condition for the supply of services (e.g., backflow-assembly testers) is unreasonably restrictive to the exclusion of qualified schools.
- A private certification program may offer a desirable program that uses third-party proctors to administer testing, thus providing a high level of quality

assurance. Other certification programs that offer basically the same quality assurance should be afforded equal consideration. There should be a rational basis for choices made in this regard, and arbitrary discrimination should be avoided.

- The water purveyor's use of a certification list from a nongovernmental organization (e.g., certification of backflow-assembly testers by AWWA's sections) is usually discretionary. However, the water purveyor should be prepared to show that its adoption of an exclusive list from that source, to the exclusion of certification lists from other sources, is reasonable and not arbitrarily discriminatory.

The water purveyor's program administrator should have related certification(s) that include

- completion of a written exam that includes questions on state or provincial regulations, water and wastewater system operations, plumbing codes, fire codes, recycled water systems, hazard assessment, basic backflow-assembly testing, and program quality assurance; and
- completion of continuing education requirements for periodic recertification.

The administrator should also have backflow-prevention assembly tester certification to demonstrate knowledge relating to field testing of backflow assemblies. The knowledge and experience required to administer a program is relative to program size, facilities, and operations encountered and type of program administered.

Minimum Standards for Acceptance of Backflow-Prevention Assemblies and Field-Test Equipment

The water purveyor must specify the minimum standards for the backflow preventers approved and relied on to protect the distribution system. This may be done by establishing an approval list that includes specific manufacturer's models for each type and size of backflow preventer, installation requirements for each listed backflow preventer, and accurate field-test equipment and calibration records. In addition, an approval list established by another authority or organization can be adopted.

State or provincial governments often publish or provide a list of approved backflow preventers or specify the approval list of an organization. Approval lists exclude products. Where the approval list used by the water purveyor is mandated by a government agency, that agency must show reasonable grounds for inclusion or exclusion of a product from the list. Where the water purveyor voluntarily adopts a list or refines a government agency's list to exclude a product or manufacturer, the water purveyor may be called on to explain the reasonableness of its action.

A list of approved backflow preventers should be based on the following:

- Manufacturer's compliance with nationally recognized standards (e.g., ANSI/AWWA C510, *Double Check Valve Backflow-Prevention Assembly*)
- Testing by an independent laboratory to show that the backflow preventer complies with the referenced standard (The testing should include a field evaluation program.)
- Inclusion of conditions for backflow-preventer installation where appropriate to performance (e.g., horizontal installation only)
- Periodic review for renewal of listing

Where a state or provincial approval list is established, the water purveyor may refine the list by establishing more stringent requirements. When doing so, the water purveyor must set clear guidelines for inclusion or exclusion from its approval list. Review by legal counsel is recommended.

To ensure the quality of the field testing of backflow preventers, the water purveyor should also specify minimum standards for the field-test equipment. This may be done by establishing an approval list or by adopting an approval list established by another authority or organization. The state or provincial government may publish a list of approved field-test equipment or specify the approval list of an organization. All field-test equipment should be checked for accuracy at least annually.

Quality Assurance

A quality assurance program is important for all aspects of operating a water system. For cross-connection control, the minimum quality assurance program should include the following:

- Review of the performance of backflow-prevention assembly testers, which includes
    - spot-checking (auditing) the tester's work by observing the backflow-prevention assembly field test or by inspecting and retesting assemblies,
    - comparing field-test data with manufacturer's data and previous field-test reports,
    - checking on the proper completion of field-test report forms,
    - verifying with the certification agency that the tester's certification is current.
- Review of the field-test results submitted for backflow preventers to determine
    - if the results are unsatisfactory (i.e., component(s) failed performance criteria);
    - that replacement or repair is needed;
    - that the backflow preventer has been replaced, relocated, repaired, modified, or removed without the water purveyor's prior knowledge; and
    - that the backflow preventer is improperly installed or in an improper application.
- Monitoring of field-test equipment to ensure that accuracy is within tolerances. This may be done by
    - requiring that field-test equipment be certified by an independent laboratory, and
    - having the water purveyor check field-test equipment.

Coordination With Local Authorities

Coordination with other authorities involved in cross-connection control may range from exchanging program information to operating a joint program. As a minimum, the water purveyor should inform local authorities (e.g., building, plumbing, and health officials) of the following:

- The water purveyor's requirements (service policies), e.g.,
    - ensuring that all premises shall be hypothetically assumed to have a check valve for the purpose of ensuring that thermal expansion is compensated for (the plumbing inspector would need to enforce the requirement for thermal expansion protection and a pressure-temperature relief valve on the water heater) or

- ensuring that all backflow-prevention assemblies installed at the service for protection or at the internal fixture in lieu of service protection must be on the water purveyor's list of approved backflow-prevention assemblies.
* The results of the water purveyor's survey of premises, including
  - the notification given to a customer of the specific requirement for service protection or internal fixture protection in lieu of service protection;
  - any survey information that indicates that the customer may not be in compliance with the plumbing code, where such omission is not a factor in determining the water purveyor's protection;
  - the notification given to a customer that the customer is in breach of the water purveyor's cross-connection control requirements (e.g., for failing to submit an annual test report); or
  - the notification given to a customer that water service will be discontinued (for cause) or other enforcement action taken.
* The receipt of water quality complaints that may indicate a cross-connection control problem.

These coordination efforts by the water purveyor are made primarily for reasons of risk and liability management. They are also beneficial for public education. Therefore, these efforts should be made unilaterally. In return, the water purveyor should request the following from the local authorities:

* that the local authority notify permit applicants that the water purveyor may have separate (additional) cross-connection control requirements;
* a copy of the results of their plan review and/or inspection of a plumbing system with regard to cross-connection control (e.g., any high hazard requiring a reduced-pressure principle backflow-prevention assembly or double check valve assembly;
* immediate notice of any enforcement action taken with regard to cross-connection control;
* immediate notice of any water quality complaint (even if it is not investigated by the local authority); and
* immediate notice of the investigation of a water quality complaint that involves cross-connection control.

As part of the water purveyor's liability management program, the request for this coordination effort by a local authority should be made in writing to the head of that authority.

Joint programs are most common when the water system is owned by a city or town that has building inspection and plumbing inspection authority. However, many large utilities are privately owned, owned by a city that has a service area outside of its city limits, or owned by a city that does not have building and plumbing inspection authority.

For any joint program, the water purveyor must

* delineate the jurisdiction of each agency;
* enter into a bilateral or multilateral agreement with the other agencies (e.g., inter-local agreement between a water district and county government) defining the responsibilities of each agency;
* designate the lead agency and combined program manager; and

- determine the procedures for all aspects of the cross-connection control program, including
  - plan review;
  - process for inspecting/surveying of premises;
  - program administration, including policy for communication with property owners;
  - enforcement of code and water purveyor's requirements;
  - record management;
  - priorities for surveying existing premises, commitment of staff time and financial resources; and
  - standards for application, installation, field testing, and repair of backflow-prevention assemblies.

Joint programs may be operated between different organizations. However, even though the water purveyor may have a joint program with another agency, the water purveyor cannot avoid the responsibility and thus liability arising from the spread of a contaminant that enters its water distribution system.

Record Keeping and Data Management

Good record keeping is essential to the proper operation of a cross-connection control program. Additionally, records may be subject to audit by state or provincial agencies. All original records (correspondence, plans, etc.) should be kept in the water system's files. If contractors are used (consultant cross-connection control specialists, backflow-prevention assembly testers), the water purveyor should retain photocopies of all records.

Record of Risk Assessment

For each customer, the water purveyor shall have a record of the initial risk assessment and subsequent reassessment, in the form of a completed water use questionnaire (residential customers) and a cross-connection survey report.

For risk management reasons, the water purveyor should retain both the initial form plus the latest reassessment, because both could

- demonstrate that the water purveyor has complied with the state/provincial requirement to evaluate new and existing customers to assess the degree of hazard;
- contain a signed statement from the customer or customer's contractor (e.g., backflow-prevention assembly tester or cross-connection control specialist) about their water use and/or assessed degree of hazard; and
- contain information useful for the investigation of a backflow incident.

Inventory of Backflow-Prevention Assemblies

For each customer where a backflow-prevention assembly or air gap is required by the water purveyor to protect its distribution system and for approved backflow preventers installed at facilities owned by the water purveyor, an inventory should include

- information on location of the backflow preventer or air gap (adequate details to locate the backflow preventers);
- description of hazard isolated; and
- type, size, make, model, and serial number of backflow preventer or air-gap details.

Backflow-Prevention Assemblies Test Reports

An inventory for each field test or air-gap inspection should include the following:
- Name and certification number of the backflow-prevention assembly tester
- Field-test results
- Repair history
- Tester's signature
- Type, size, make, model, and serial number of backflow preventer or air-gap details
- Positive identification of the assembly, which may include field tagging

Correspondence

The water purveyor should maintain all correspondence with its customers for at least 3 to 5 years or as required by local jurisdiction. The most current service agreement and instructions to install backflow preventers to protect the water purveyor's distribution system should be maintained as a permanent record. All correspondence with the state or provincial authority and the local administrative authority should be maintained for at least 5 years.

Spreadsheets and Computer Database Software

Based on the number of backflow-prevention assemblies present in most small water systems, the inventory of backflow-prevention assemblies, test reports, local backflow-prevention assembly testers, etc., can be kept on a spreadsheet, in either paper form or on computer software. Proprietary computer database programs developed specifically for managing a cross-connection control program are available. The size of the cross-connection control program must be considered when determining specific data management requirements.

Backflow Incident Reports

Details on the investigation and subsequent corrective action taken for reported backflow incidents should be kept. Some states and provinces require that details of incidents be submitted to them. They may have a required backflow incident form that must be completed.

Public Relations and Education

Success in operating a cross-connection control program is often linked to the public's understanding of the public health hazards posed by cross-connections. No one would connect plumbing fixtures, equipment, etc., to their drinking water system knowing it could pollute and/or contaminate their drinking water.

A public education program is an integral part of a cross-connection control program. Public education programs vary based on the water purveyor's resources and the types of customers served. For all education efforts, the following information should be conveyed:

- The nature of the public health risk posed by actual or potential cross-connection hazards
- The fact that the water purveyor is responsible for protecting the public water system from contamination

- The fact that the customer is responsible for preventing a contaminant from entering their plumbing system and thereafter entering the public water system
- The fact that the water purveyor is required to comply with state or provincial regulations concerning cross-connection control
- The fact that the water purveyor has established policies (conditions of service) relating to cross-connection control

The audience. Public education efforts should be tailored to the intended audience. Following are examples for two major categories.

Single-family residential customers. For single-family residential customers (i.e., the general public), educational materials must explain what constitutes a cross-connection, how backflow can occur, etc., in easy-to-understand, nontechnical terms. The material should be kept to a reasonable length and use graphics to illustrate key points and keep the reader interested.

Common forms of public education for this group of customers include

- water bill inserts (brochures);
- Consumer Confidence Reports;
- information on the service policy distributed with application for service for new customers;
- mall, fair, home show displays, and similar venues; and
- use of the news media.

The annual Consumer Confidence Report should contain a brief statement describing the water purveyor's operation of a cross-connection control program and the customer's responsibility to protect their plumbing system and the water purveyor's distribution system. Educational brochures should be distributed every 2 to 3 years.

Public service announcements on radio and television are very effective. Their announcements are generally free, but there are costs for developing a short video(s) for television. Newspaper articles can also be used. Alerting the consumer to the dangers of unprotected lawn sprinkler systems, using a hose to spray weed killer and fertilizers, or unplugging drains are examples of article topics.

Commercial customers. This group includes all customers other than the single-family residential group. The educational information provided to commercial customers may be more technical. It should be directed to the specific customer (e.g., dry cleaner, dental clinic, shopping mall owner, restaurant, etc.). Commercial customers are likely to confer with their maintenance staff, plumber, architect, etc., regarding the information provided. In many cases, the cost to the commercial customer of installing backflow preventers on existing premises is significant compared to their operating budget. This may generate complaints or other resistance regarding compliance that must be addressed with educational materials.

Professional and trade groups. When providing education to technical groups, the utility should stress that customers rely on their consultants, contractors, etc., to fulfill the customer's responsibility to prevent contamination. Examples of technical groups include

- Occupational Safety and Health Administration inspectors,
- health inspectors,
- architects and engineers,
- building and/or plumbing inspectors,
- plumbing suppliers,
- plumbing contractors and plumbers,

- wastewater treatment personnel,
- irrigation contractors and suppliers,
- fire suppression contractors,
- pool and spa contractors,
- pest control product suppliers, and
- pipe fitters and mechanical contractors.

Smaller utilities located near large utilities with good cross-connection control programs can join forces with the larger ones. In some areas, this can be done through local AWWA sections or by establishing a local cross-connection control committee. These efforts promote good public relations. It is better to use persuasion instead of enforcement of regulations to obtain public health protection. Willing customers are more likely to maintain their backflow preventers and are more forgiving of the water purveyor for imposing the added costs of backflow prevention.

Safety

Operation of a cross-connection control program may involve the water purveyor's staff who will install, test, and maintain backflow preventers and survey customers' premises for cross-connections. Consequently, the water purveyor's safety program should recognize the following issues, which relate specifically to cross-connection control:

- Repair of a backflow-prevention assembly often requires special tools. Use of the wrong tool may result in injuries to the worker or damage to the backflow-prevention assembly. For example, metal tools used to clean seats may damage them, and large assemblies may have heavy covers, and their springs may be under great tension. Additionally, the need for disinfection should be evaluated.
- Many backflow-prevention assemblies are installed in hazardous locations, particularly old assemblies installed on the customer's premises. The potential hazards related to these assemblies include confined spaces, near ceilings requiring platforms, exposure to hazardous or toxic materials in industrial plants, and automobile traffic or moving equipment.

Backflow Incident Response Plan

Administration of a cross-connection control program should include the investigation of any water quality complaint that indicates possible backflow contamination. Quick response by personnel trained in cross-connection control and basic water quality is necessary for the success of any investigation of a backflow incident and prevention of further contamination. The water utility should act in cooperation with other authorities or agencies, such as the health inspector, to protect the consumers of water within the premises where a contaminant has been detected.

There are numerous well-documented cases of cross-connections causing contamination of drinking water. An overview of the potential for contamination, the difficulties in identifying the source of contamination, and the efforts to remove a contaminant from a water system can be obtained by reviewing published backflow incidents (see appendix F).

The water purveyor's investigative actions are intended to

- protect the water purveyor's distribution system from the spread of a contaminant detected in the water supply on private property,

- quickly restore the quality of water in the utility's distribution system if a contaminant has entered the system through backflow from the customer's plumbing system, and
- prevent any further contamination of the utility's system.

Written guidelines on how to respond to all foreseeable complaints should be included in the water purveyor's management and operations manual. Even when written guidelines are provided, it is impossible to cover all types of water quality complaints and other operating scenarios within the water purveyor's system that may contribute to a water quality complaint.

Most backflow incidents such as the backflow of "used" water, water with a low level of chemical contamination, or water with an undetectable bacteriological contaminant are not likely to be identified as such. A person may complain to others, but not to the water purveyor, about water with a slight taste or odor. When a complaint is made to a utility, the utility may not respond expeditiously and thus may not find conclusive evidence that a contaminant has entered the potable water system through a cross-connection.

When the initial evaluation of a water quality complaint indicates that a backflow incident has occurred (potable water supply has been contaminated/polluted) or may have occurred or when the reason for the complaint cannot be explained as a "normal" aesthetic problem, a backflow incident investigation should immediately be initiated. It is wise to be conservative when dealing with public health matters.

A backflow incident investigation is often a team effort. The investigation should be made or (initially) led by the water purveyor's cross-connection control program administrator. If the administrator does not have training in basic water quality monitoring, a water quality technician should be part of the investigation team. One or more water distribution system operators may be called on to assist in the investigation and to take corrective action. The investigation team may also include local health and plumbing inspectors.

The water purveyor is involved in any backflow incident investigation because they received the complaint and have an obligation to make a reasonable response and because they have an interest in preventing the spread of a contaminant through their water distribution system.

Each investigation is unique and requires the development of a specific plan of action at the start of the investigation, with modification to the plan as the investigation progresses.

The water purveyor's investigation should include these steps:

1. Locate the source of the contamination.
2. Isolate that source to protect the water distribution system from further contamination.
3. Determine the extent of the spread of contamination through the distribution system and provide timely, appropriate notification to the public and to regulatory agencies.
4. Take corrective action to clean the contamination from the distribution system.
5. Restore service to the customers.

The traditional cure for distribution system contamination, flushing by the water purveyor, must be thoroughly evaluated before being implemented because flushing can, in some cases, contaminate more of the system.

The public health authority must consider the needs of those who may have consumed or used the contaminated water. And the customer must take any action necessary to clean the plumbing system, shut down equipment, vacate the premises, and similar steps.

In any circumstance where a claim for damages may follow, the water purveyor should establish (with the advice of their attorney or risk manager) the procedures to be followed to provide advice or assistance to the customer. Additionally, the water purveyor should complete a backflow incident investigation report.

The water purveyor may need to notify the public of contamination of the distribution system. A public notification procedure should be part of the water purveyor's general emergency plan. Failure to provide timely, appropriate notification to all customers that may be affected by water system contamination may result in a claim for damages by customers. Appropriate notification procedures are essential for liability management.

The followup to a backflow incident investigation should include debriefing of personnel, reviewing response and investigation procedures, and accounting for the cost of the investigation.

Budget and Sources of Funding

The cost of the cross-connection control program should be identified in the water purveyor's operating budget. To reduce and/or recover the cost of the program, the water purveyor may assess the general administrative cost to all customers or to one class of customer (e.g., all commercial customers) through use of

- commodity or consumption charges ($/gallon);
- meter charges ($/month), based on size of meter;
- supplemental backflow-prevention assembly charge ($/month), based on size of assembly; or
- charge for installation permits.

The water purveyor may also require each customer to directly bear the survey, purchase, installation, testing, and maintenance costs of backflow-prevention assemblies.

This page intentionally blank.

AWWA MANUAL M14

Chapter 3

Backflow Principles

BASIC HYDRAULICS

To help the reader understand the hydraulic conditions that may result in the backflow of a contaminant into a potable water system, this section defines and illustrates basic terms.

Water pressure. Water pressure is defined as the amount of force acting (or pushing) on a unit area. Pressure is created by pumping, compression, or elevation. It is defined as force per unit area and is most commonly expressed as pounds per square inch (psi). Water pressure is directly related to the elevation of the water. Based on the fact that water has a density of 62.4 lbm/ft^3 (999.6 kg/m^3), the force of 1 ft^3 of water pushing down on a 1-ft^2 (0.09-m^2) surface area is 62.4 lb (277.6 N). In converting this to pounds per square inch, a 1-ft (0.3-m)-high column of water pushing down on a 1-in.^{2} (645.2-mm^2) surface area weighs 0.433 lb (1.9 N), resulting in a pressure of 0.433 psi (0.003 N/mm^2, or 3 kPa). Because water pressure is directly related to the height of water, pressure also can be stated as feet of head. One foot of head equals 0.433 psi (2.3 kPa).

Atmospheric pressure. The earth is surrounded by atmospheric gases. Because of the gravitational pull of the earth, these gases exert a pressure of 14.7 psi (101.4 kPa) at sea level. There are two references from which fluid pressures are measured: absolute pressure and gauge pressure. Absolute pressure is the measure of the combined force of gauge pressure and atmospheric pressure. Gauge pressure is the measure of force above atmospheric pressure. In the United States, pressure usually is measured in pounds per square inch. To distinguish between the two pressure references, the letters *a* and *g* often are added as suffixes to the abbreviation psi, hence, *psia* and *psig*. A standard atmosphere at sea level is 14.7 psia (101.4 kPa); thus, an open system under 1 standard atmosphere and no flow conditions is at 14.7 psia (101.4 kPa), or 0 psig. An easy distinction can be made between atmospheric and gauge pressure: Gauge pressure is simply the pressure in excess of atmospheric pressure. It is important to understand the influence of atmospheric and negative gauge pressure on unprotected cross-connections and backflow.

Vacuum. A vacuum occurs when a condition of negative gauge pressure (psig) exists. A vacuum creates a suction, and liquids always flow to the area of least resistance or lowest pressure. Because atmospheric pressure at sea level exerts 14.7 psia (101.4 kPa) on the surface of an open container of liquid, one can calculate the height to which a liquid would rise in a column where a perfect vacuum is created (see Figure 3-1). At sea level, the atmospheric pressure (14.7 psia [101.4 kPa]) would push the fluid up the column 33.9 ft (14.7 psi × 2.31 ft/psi = 33.9 ft). Therefore, the maximum height that water can be "lifted" with a perfect vacuum is 33.9 ft (10.3 m).

Barometric loop. A barometric loop is based on the principle that at sea level a column of water will not rise or be siphoned above 33.9 ft (10.3 m). A barometric loop consists of a continuous section of piping that vertically rises to a height of approximately 35 ft (10.6 m) and then loops down to return to the original level. This loop in the piping system effectively protects against backsiphonage but not against backpressure. Check with the jurisdiction that has authority to determine whether the barometric loop is an approved method of backflow protection in any particular area.

Venturi effect. A venturi effect is created when there is a local restriction in the water line. If water is forced through a restriction, the velocity must increase. As velocity increases, the pressure must decrease. A connection at the point of the restriction could create a suction, or backsiphonage condition. This principle frequently is used to intentionally introduce chemicals, such as chlorine, into a water line. Chemical sprayers that attach to the end of a garden hose, used to apply fertilizers or pesticides to lawns, typically use the venturi principle to siphon the material from the container.

Hydraulic grade line. This is a theoretical line created to determine the pressure at any point in the water distribution system, taking into account the elevation differences and friction losses. The line is determined by calculating the vertical rise of water at various points in the distribution system. To adequately maintain system pressure, it is critical that all service-connection water demands in the system fall below this line under all flow conditions. Backsiphonage conditions will occur

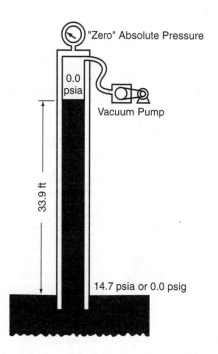

Source: US Environmental Protection Agency. Cross Connection Control Manual *(1989).*

Figure 3-1 Schematic of a vacuum

at any point where the water demand exceeds this line under any flow condition, typically during periods of high demand or flow. Water distribution systems with greater elevation differences are more susceptible to backsiphonage conditions at the higher elevations.

TYPES OF BACKFLOW

Backflow events can result from backsiphonage or backpressure.

Backsiphonage Backflow

A backsiphonage condition is created when a subatmospheric pressure is applied to the inlet of the piping system. Whenever the pressure in the piping system drops below atmospheric pressure, negative pressure is created. Because negative pressure in the water supply system can cause backflow, it is important to understand the conditions or actions that can create negative pressure. These conditions can be illustrated by calculating the hydraulic grade line of a public water system and comparing it with all of the known water uses during all flow conditions.

To prevent any backflow into the public distribution system, the water purveyor should maintain a minimum pressure at all points in the distribution system under all flow conditions. Many states have regulations stating minimum pressure requirements.

Typical conditions or arrangements that may cause backsiphonage conditions include high-demand conditions (fire flow, customer demand during heat-wave emergencies), inadequate public water system source and/or storage capacity, and water main breaks.

Figure 3-2 shows how water main pressure is affected if water is withdrawn at normal and high rates. Under normal flow conditions, all service connections fall below the hydraulic grade line. Assume the hydrant at point F is opened during a period of high demand and the valve at point G has been closed, restricting the supply of water to the area or hydrant. The pressure at the hydrant drops, leaving the storage tank at point B, the top floors at point C, the house and swimming pool at point D, and the house at point E above the hydraulic grade line. The pressure is now reduced to a point where water can no longer be supplied to these areas. To equalize the pressure, water in the lines in these areas will flow toward the lower pressures, thus creating backflow conditions.

The backsiphonage shown in Figure 3-3 is caused by reduced water system pressure on the suction side of an online booster pump. The pressure of the public water main is adequate to supply water only to the first and second floors; therefore, a booster pump is required to service the upper floors. The potable water supply to the dishwasher on the second floor is not protected by a backflow-prevention assembly and has a direct connection to the sewer. When periods of high demand coincide with periods of low pressure in the public water system, the booster pump that supplies the upper floors could create a backsiphonage condition by further reducing the pressure in the service connection that supplies the lower floors of the building.

The backsiphonage depicted in Figure 3-4 shows that when a distribution system is shut down to accommodate a repair, negative or reduced pressure will occur at all locations of the affected system that are located at any elevation higher than the break. The water main break in the street causes negative pressure in the house; this, in turn, causes contaminated water to be drawn from the bathtub back toward the main. This type of backsiphonage condition usually affects more than one service connection; multiple city blocks can be affected, including commercial and industrial connections as well as residential connections.

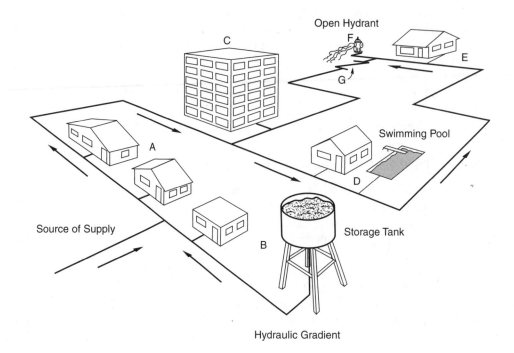

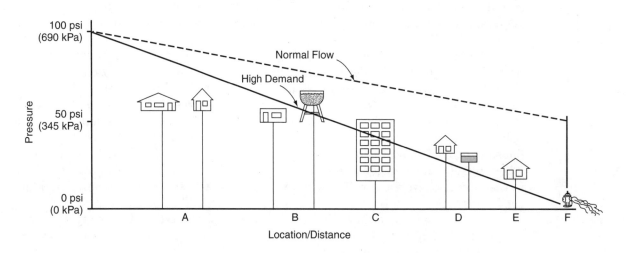

Figure 3-2 Backsiphonage backflow due to high rate of water withdrawal

Backpressure Backflow

Backpressure is the condition that can occur when the pressure on the downstream side of a piping system is greater than the pressure on the upstream side. When the downstream side of the piping system is connected to a nonpotable source, unacceptable backflow could occur.

Common causes or sources of backpressure include

- pumps,
- elevated piping,

BACKFLOW PRINCIPLES   33

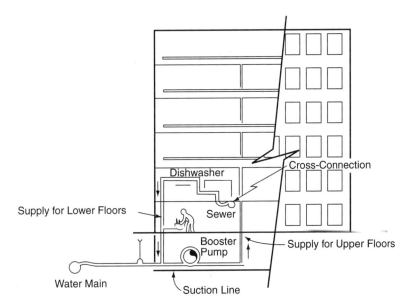

Figure 3-3 Backsiphonage backflow caused by reduced pressure on suction side of booster pump

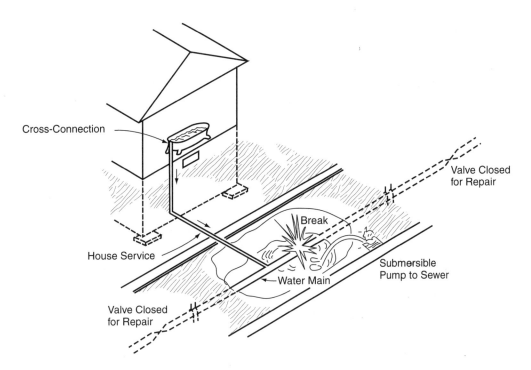

Figure 3-4 Backsiphonage backflow caused by shutdown of water system

- thermal expansion,
- pressurized containers.

Booster pumps are commonly used to meet fire demands or manufacturing-process demands. They may also be used for chemical-feed systems, auxiliary irrigation systems, car washes, and cooling systems. If a pump is used on a property served by a public water system, there is a possibility that the pressure downstream of the pump may be higher than the pressure in the potable water system.

Elevated piping. Whenever a potable water system serves water to facilities that are located at elevations above the water system's source, pressure from the weight of the column of water creates backpressure; this may produce backflow. Examples of elevated piping include water storage tanks, fire sprinkler systems, and high-rise buildings.

Thermal expansion. Thermal expansion is a physical property: All liquids and gases expand when heated. The expansion created when water is heated in a closed container may produce pressures greater than the water supply pressure; this results in backpressure. To prevent backflow, a closed system must have a means to safely absorb or relieve the effects of excessive pressure caused from thermal expansion.

Boilers are a common source of backpressure backflow caused by thermal expansion. Boiler makeup water typically is required for all boilers, and chemical additives may be added to reduce scale and corrosion. Whether or not chemical additives are used, makeup water must be isolated from the potable water system to prevent a backflow hazard. Another source of thermal expansion is fire sprinkler systems that are located at high points of buildings and are subject to the ambient air temperature. As the air temperature increases, the water expands and the pressure in the piping system increases.

Pressurized containers. Gases under pressure are used in various commercial applications. One of the most widespread uses of gas is the supply of carbon dioxide for carbonating beverages. Carbon dioxide cylinders used for post-mix beverage-dispensing machines are found throughout the food service industry. (Figure 3-5 illustrates a plumbing connection commonly used for post-mix beverage-dispensing machines.)

Pressurized tanks are also used in many industrial facilities. For example, hydropneumatic tanks are used on pumping systems to protect motors from frequent starts and stops. A hydropneumatic tank is partially filled with water or other liquid, and the rest of the tank is filled with air or another gas. The compressibility of the gas allows the liquid to be supplied within a desired pressure range without pumping.

Backflow is possible when the water purveyor's supply pressure is less than the pressure of the gas cylinder or hydropneumatic tank. The volume of liquid stored in the hydropneumatic tank adds to the volume of liquid in the piping system that could flow back into the potable water system.

Figure 3-6 shows how a pump on the customer's water system can increase the water pressure to a point where it exceeds the public water distribution system pressure, causing a backflow condition. It is a common practice to flush ships' fire-fighting systems by connecting them to dockside freshwater supplies. As shown on the graph, under normal conditions the pressure in the main is 100 psi (689.5 kPa), and it is approximately 75 psi (517.1 kPa) where it enters the ship's system. After completing the flushing operation, a test is conducted to determine whether the fire pumps aboard the ship are operating properly. As shown on the graph, the fire-system pressure is boosted to 200 psi (1,379.0 kPa). If the valve at point A is accidentally left open, the fire-system pressure, which is higher than the public water distribution system pressure, forces salt water into the dockside and public water systems.

ASSESSING DEGREES OF HAZARD

Properly applying a backflow preventer depends on an accurate assessment of the risk of contamination of a potable water supply. This applies to a single cross-connection (i.e., a plumbing fixture) or a group of cross-connections (i.e., an entire plumbing system).

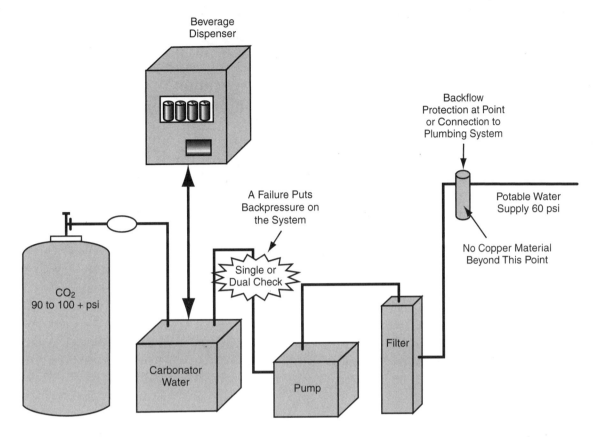

Figure 3-5 Backpressure backflow caused by carbon dioxide cylinder

The importance of accurate risk assessment is reflected in a common phrase used in reference manuals and regulations: "a backflow-prevention assembly commensurate with (appropriate for) the degree of hazard shall be installed …."

However, risk assessment is subjective. A risk that is considered appropriate by one person or group may be considered inadequate or excessive by another. There also is disagreement about assessing risk in relative terms, e.g., whether a cross-connection poses a high or low hazard, whether one type of cross-connection poses a higher hazard than another type does.

Assessing the Risks of Cross-Connections

Every cross-connection poses a different risk based on the probability of

- the occurrence of a physical connection between a potable water supply and a nonpotable substance;
- the occurrence of backflow conditions (e.g., a water main break that causes backsiphonage);
- the total failure (however unlikely) of the backflow preventer used to isolate the cross-connection; and
- the probability that a nonpotable substance is present and will have an adverse effect on the water system or user.

For a potable water system to become contaminated, all four factors must be simultaneously present.

36 BACKFLOW PREVENTION AND CROSS-CONNECTION CONTROL

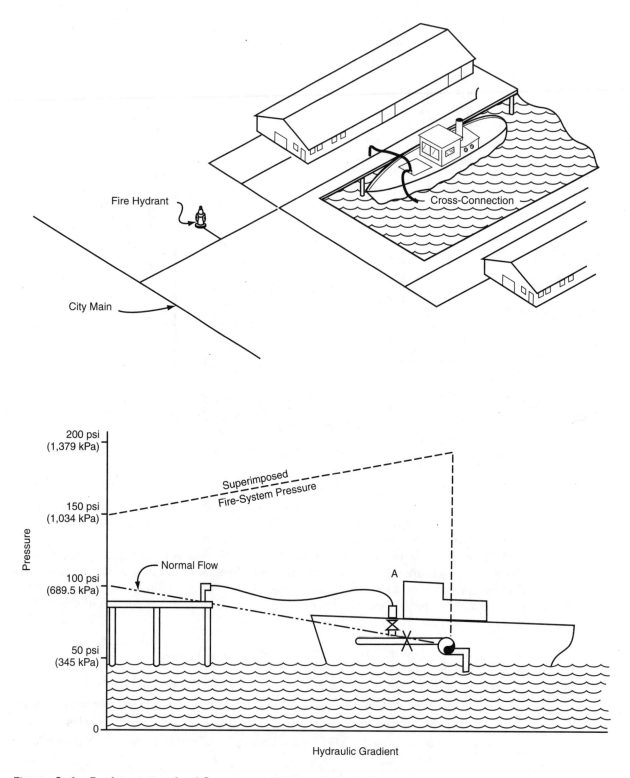

Figure 3-6 Backpressure backflow caused by pumping system

Actual Connections and Potential Connections

The probability of an actual connection is a certainty. Actual connections are common. Examples include a submerged connection of a potable water line made to a plating tank and a connection of a potable water line made to a lawn irrigation system.

In the absence of statistical information, estimating the likelihood that a potential connection will become an actual connection relies upon the experience and specialized training of the individual making the assessment. Reviewing reports about backflow incidents provides information about factors leading to the creation of actual connections and accompanying backflow conditions.

There are multitudes of potential connections. Examples of potential connections are a hose-bib outlet located near a plating tank and an irrigation system supplied from a pond or other auxiliary supply located on premises that are supplied with potable water service.

Without statistical information, it is not possible to estimate the probability that any given potential connection will become an actual connection. Because of the multitude of potential connections, it is not practical to collect statistical information with an adequate sampling to be valid.

One difficulty in assessing the probability of a potential connection becoming an actual connection relates to changes made to plumbing in customers' premises. For example, do-it-yourself homeowners can easily modify residential plumbing systems. Homeowners often purchase and install lawn sprinkler systems, residential solar heating systems, etc., without obtaining plumbing permits. Similarly, in any complex piping system (e.g., industrial plant, hospital) there is an inherent risk that maintenance personnel or other persons may modify the plumbing system.

Assessing the Risk of Backflow Conditions

In contrast, it is much easier to assess the probability of occurrences of backflow conditions (backpressure or backsiphonage) than it is to assess the probability of potential and actual connections.

Backpressure conditions may be evaluated as actual or potential conditions. Examples of actual backpressure conditions include a connection made to a high-pressure heating boiler and a connection made to a booster pump that supplies water to a high-rise building. Examples of potential backpressure conditions include devices, such as a standpipe connection or sprinkler connection, that are installed for use by the fire department to pump additional water into a fire sprinkler system and a connection made to a hot-water tank. An increase in pressure may occur if the tank's pressure-temperature relief valve fails.

Backsiphonage is evaluated primarily as a potential condition. The following are examples of potential backsiphonage conditions:

- Water main breaks in a distribution system. The greater the number of breaks (say, per mile of pipe), the greater the frequency of backsiphonage conditions.
- The prevalence of hilly terrain. Any water main break, fire flow, closed valve, or like condition increases the likelihood that backsiphonage conditions will occur on hilltops.
- Limited hydraulic capacity and redundant components in the supply and distribution. A major reduction or loss of positive pressure may occur during high-flow conditions (e.g., peak-hour demand plus fire flow). Shutting down a major system component (e.g., a transmission line or booster pump station)

may result in a major reduction or loss of positive pressure in the distribution system, increasing the likelihood of backsiphonage conditions.

It is comparatively easy to assess the relative probability that backsiphonage conditions will occur. Many of the underlying causes of backsiphonage conditions are under the control or management of the water purveyor and, thus, are well understood.

Assessing Other Risk Factors

Assessing the effectiveness of backflow preventers is discussed later in this chapter.

The public health aspects of cross-connection control were discussed in chapter 2. In general, the public health risk of a nonpotable substance ranges in descending order from acute microbial to acute chemical to chronic chemical. Classifying the risk of the many microbial and chemical contaminants is not practical. Therefore, the possible addition of any substance to potable water poses a risk to the water purveyor's system. For example, even where a nontoxic chemical is added to a plumbing system, there is a risk that the chemical will be unintentionally replaced by a toxic chemical.

Furthermore, any water that leaves the control of the water purveyor should be considered at risk of contamination. Potable water quality does not improve with age, particularly when it remains stagnant in a customer's plumbing system. For this reason, the AWWA policy statement (which appears in appendix D) opposes the return of "used water" to the water purveyor's system.

Putting It All Together

The probability of a number of independent events occurring simultaneously equals the product of the probability of each event occurring. For example, if the probability that a backflow-prevention assembly will fail is 20% ($1/5$), and the probability that backsiphonage will occur from a shutdown of supply resulting from a power failure is 10% ($1/10$), then the probability that a backflow-prevention assembly will fail at the same time that backsiphonage will occur is 2% ($1/5 \times 1/10 = 1/50$).

Unfortunately, it is difficult to determine the probability of occurrence of the many events that influence the risk of contamination. For this reason, selecting a backflow preventer "commensurate with the degree of hazard" for fixture or service containment remains a subjective decision. However, it is clear that the lack of a backflow preventer significantly increases the potential that backflow will occur; and the greater the effectiveness of the backflow preventer, the lower the potential that backflow will occur.

From the water purveyor's perspective, each of its customers' plumbing systems poses a potential health hazard. The degree of hazard ranges from low to high. A "non-health" hazard assessment is inappropriate. In assessing the degree of hazard, the water purveyor must focus on the overall hazard posed by a customer's entire plumbing system. For example, for a chemical contaminant, the hazard at the fixture may be significantly greater than the hazard to the distribution system, because the concentration of the chemical is higher at the fixture. Chapter 5 provides guidance about typical hazards posed by various categories of customers.

ASSESSING THE EFFECTIVENESS OF ASSEMBLIES AND DEVICES

Any means designed to stop backflow, when it is working properly, can be considered a backflow preventer. These may range from a dual check valve to a backflow-prevention assembly. However, not all backflow preventers are considered equal in stopping backflow. The common types of backflow preventers are described in chapter 4.

Any mechanical apparatus can fail to perform as designed. Failure may be the result of a design flaw, operating conditions that exceed design parameters, improper installation, normal wear on moving parts, corrosion, etc. Factors that increase the reliability of a backflow preventer include

- Construction to meet design and performance standards.
- Verification of compliance through evaluation by an independent testing agency that has experience in evaluating backflow preventers.
- Inclusion of a field evaluation as part of the verification of compliance.
- Accelerated life-cycle tests.
- Periodic review of a testing agency's approvals to ensure that its processes meet the water purveyor's needs.
- Provision in the design standard for a method of testing a backflow preventer to confirm that it meets performance criteria.
- Provision in the design standard for a method of field testing and repairing a backflow preventer in place. (Removing a backflow preventer from a line increases the likelihood that a spool piece will be installed.)
- Use of field-testing methods (e.g., testing a check valve in the direction of flow) that indicate substandard performance (i.e., failure to meet performance criteria) before a component actually fails.
- Frequent testing of assemblies (e.g., a minimum of once per year) to determine the need for repair or replacement.
- Assurance (through regulation) that assemblies will be maintained, repaired, or replaced if test results indicate the need to do so.
- Assurance that testing and maintenance is performed by qualified (trained) personnel.
- Assurance that persons testing backflow preventers are qualified through certification programs and audits of their work.
- Correct installation of backflow preventers to ensure proper operation and to facilitate testing.

The highest degree of protection from backflow is an approved air gap. However, air gaps can easily be eliminated. In addition, they can be dangerous if they are located where they could be exposed to toxic fumes.

The reliability of backflow preventers that are designed for in-line testing and maintenance can be confirmed. Indeed, testing to verify continuing satisfactory performance is key to an effective cross-connection control program. Generally, assemblies are backflow preventers that require certain parts (such as test cocks and shutoff valves) that allow field testing. Assemblies must be able to be tested and repaired in line. They must meet an approval standard based on performance and design. Backflow devices, on the other hand, are not always designed to allow field testing. The approval standards for devices differ and do not provide the same level of protection and performance that assemblies provide. Definitions of assemblies and devices are included in the glossary; chapter 4 discusses common types of assemblies and devices.

The effectiveness of backflow preventers can only be generally quantified. The relative effectiveness of backflow assemblies can be established by a review of field-test

Table 3-1 Means of backflow prevention

	Degree of Hazard			
	Low Hazard		High Hazard	
Means	Back-siphonage	Back-pressure	Back-siphonage	Back-pressure
Air gap (AG)	X		X	
Atmospheric vacuum breaker (AVB)	X		X	
Spill-resistant pressure-type vacuum-breaker assembly (SVB)	X		X	
Double check valve assembly (DC or DCVA)	X	X		
Pressure vacuum-breaker assembly (PVB)	X		X	
Reduced-pressure principle assembly	X	X	X	X
Reduced-pressure principle detector assembly	X	X	X	X
Double check valve detector check assembly	X	X		
Dual check device (internal protection only)	X	X		
Dual check with atmospheric vent device (internal protection only)	X	X		

results. Generally, the appropriate application of and the relative effectiveness provided by backflow preventers can be assessed as listed in Table 3-1.

This table points out practical applications of technology. For protection of the water purveyor's distribution system through a service protection policy, the choice is normally between the installation of a reduced-pressure backflow assembly and a double check valve assembly.

AWWA MANUAL M14

Chapter 4

Backflow-Prevention Assembly Application, Installation, and Maintenance

Selecting the correct backflow preventer requires thorough knowledge of several variables. The type of backflow (backpressure and/or backsiphonage) must be identified. The degree of hazard (high hazard or low hazard) also must be identified. In addition, installation criteria and hydraulic conditions must be evaluated before proper selection of a backflow preventer. All of these criteria must be evaluated by trained personnel before a backflow preventer can be installed.

A comparison of any two manufacturers' products may reveal that the products prevent backflow in different ways. To ensure that products offer an acceptable level of protection, several independent organizations have developed and use design and performance standards for backflow preventers. These standards can vary in their requirements that a backflow preventer must meet. It is important that backflow preventers meet an acceptable standard and that they are evaluated to confirm they are approved and meet the requirements of the standard. In addition, it is important that the local administrative authority review the various assembly standards and select a standard that provides backflow preventers that will provide continual and proper protection.

There are two categories of mechanical backflow preventers to choose from: assemblies and devices. Assemblies are backflow preventers that are required to have certain parts, such as test cocks and shutoff valves, that are used for field testing. Assemblies must be able to be tested and repaired in-line. They must meet an approval standard for performance and design. Backflow devices are not always designed for field testing. Standards for devices and assemblies differ, with various standards describing different performance requirements.

After a means of preventing backflow is selected, it is critical that the backflow preventer continue to work as designed. To ensure this, a field-testing and repair protocol must be followed. Backflow preventers must be periodically tested in the field to determine whether they continue to prevent backflow properly. In addition, the backflow-prevention installation can be evaluated for modifications, changes in plumbing, and correct hazard application.

BACKFLOW CONTROL METHODS

The following sections describe applications of technology intended to prevent backflow. The functional capabilities of the installations or equipment and the recommended applications are also described.

Air Gap

Description. A proper air gap (AG) is an acceptable method to prevent backflow. The disadvantage of an air gap is that supply-system pressure is lost. An approved air gap is a piping arrangement that provides an unobstructed vertical distance through free atmosphere between the lowest point of a water supply outlet and the overflow rim of an open, nonpressurized receiving vessel into which the outlet discharges. These vertical physical separations must be at least twice the effective opening (inside diameter) of the water supply outlet but never less than 1 in. (25 mm). In locations where the outlet discharges within three times the inside diameter of the pipe from a single wall or other obstruction, the air gap must be increased to three times the effective opening but never less than 1.5 in. (38 mm). In locations where the outlet discharges within four times the inside diameter of the pipe from two intersecting walls, the air gap must be increased to four times the effective opening but never less than 2 in. (51 mm). Air gaps should not be approved for locations where there is potential for the atmosphere around the air gap to be contaminated. Nor should the inlet pipe be in contact with a contaminated surface or material.

Application. An air gap can be used for service or internal protection (Figures 4-1 through 4-4). A properly installed and maintained air gap is the best means

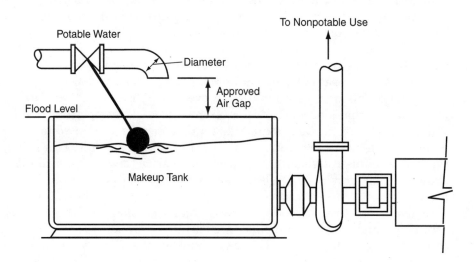

Figure 4-1 Air gap on tank

BACKFLOW-PREVENTION ASSEMBLY APPLICATION, INSTALLATION, AND MAINTENANCE 43

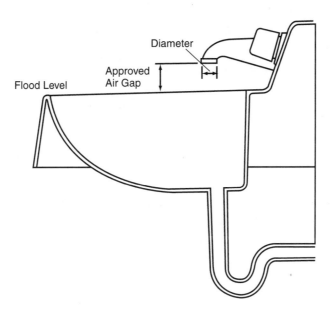

Figure 4-2 Air gap on lavatory

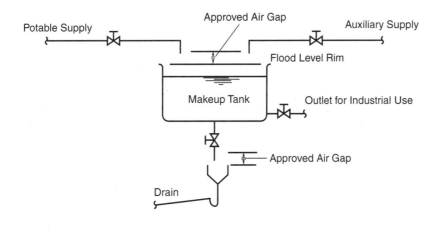

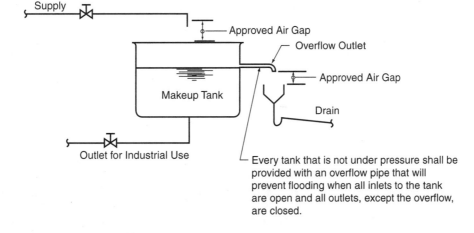

Figure 4-3 Typical air-gap applications

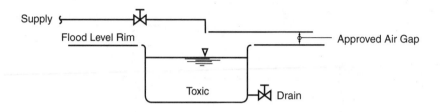

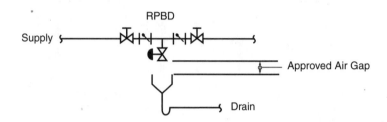

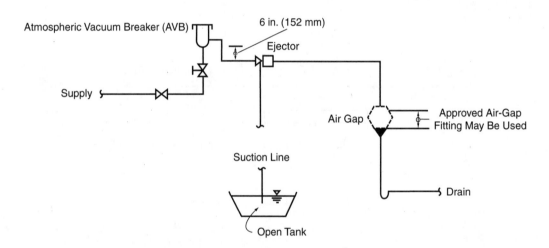

Figure 4-4　Additional typical air-gap applications

to protect against backflow because it provides a physical separation between the water source and its use. Because, in a backsiphonage condition, an air gap can allow the surrounding atmosphere to enter the piping system, care must be taken to locate the air gap in such a way to ensure that fumes or other airborne substances cannot be siphoned into the potable water system. The installation shall not include any interference with the free-flowing discharge into the receiving vessel. This means no solid material shields or splash protectors can be installed. Screen or other perforated material may be used if it presents no interference. Air gaps must be inspected at least annually to ensure that the proper installation is maintained and has not been circumvented.

Reduced-Pressure Principle Backflow-Prevention Assembly

Description.　A reduced-pressure principle backflow-prevention assembly (RP) is an assembly that shall contain two loaded, independently acting check valves with a hydraulically operating, mechanically independent pressure-differential relief valve

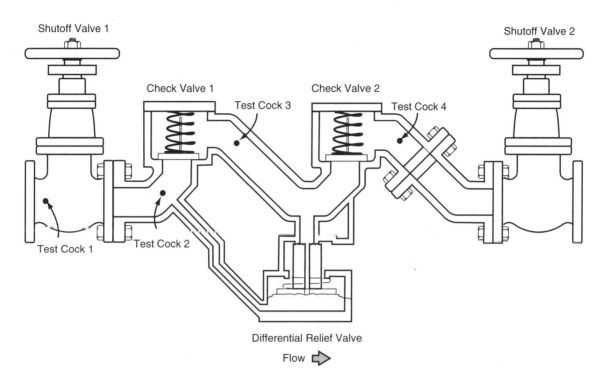

Figure 4-5   Reduced-pressure principle backflow-prevention assembly

located between the check valves below the first check valve. The check valves and the relief valve shall be located between two tightly closing, fully ported, resilient-seated shutoff valves. The RP shall have four properly located resilient-seated test cocks, as shown in Figure 4-5. The RP shall be installed as an assembly as designed and constructed by the manufacturer. An RP shall be approved by an approval agency acceptable to the local administrative authority.

Function. An RP is designed to maintain a pressure that is lower after the first check than it is at the inlet. The water pressure into the RP will be reduced by the amount of the first check loading (a minimum of 3 psi [20.7 kPa] higher than the relief-valve opening point). When the pressure is reduced after the first check, the relief valve senses the difference between the inlet pressure (before the first check) and the pressure after the first check. The relief valve ensures that the pressure after the first check is always lower than the inlet pressure by the amount of the relief-valve opening point, which shall be a minimum of 2 psig (13.8 kPa gauge). If the pressure in the area after the first check increases to within a minimum of 2 psig (13.8 kPa gauge) less than the inlet pressure, the relief valve will open to ensure that a lower pressure is maintained. The second check is located downstream from this relief valve. The second check will also reduce the pressure by the amount of the check loading, which is a minimum of 1 psi (6.9 kPa). In a normal flowing situation, both check valves will be open to meet the demand for water and the relief valve will stay closed. When the demand for water ceases, both checks will close and the relief valve will stay closed. See Figures 4-6 and 4-7.

In a backpressure condition, both check valves will close, and the second check will stop the increased pressure from traveling into the area between the two checks (Figure 4-8). If the second check were not maintaining its separation of pressure, the backpressure would leak past the second check and cause the pressure in the area between the two checks to increase. After the increase in pressure rises to the inlet

46 BACKFLOW PREVENTION AND CROSS-CONNECTION CONTROL

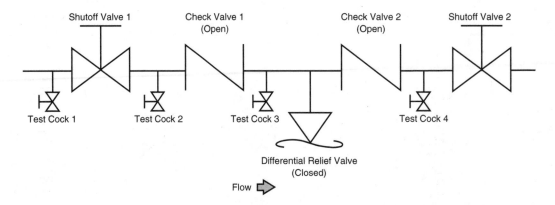

Figure 4-6 Both check valves open and the differential relief valve closed

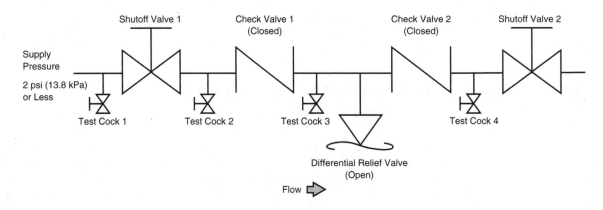

Figure 4-7 Both check valves closed and the differential relief valve open

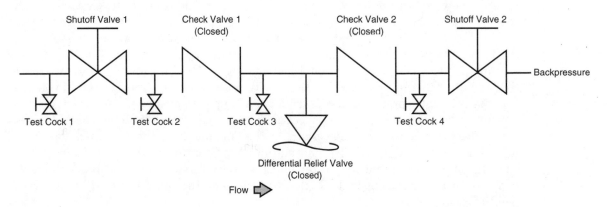

Figure 4-8 Backpressure: both check valves closed and the differential relief valve closed

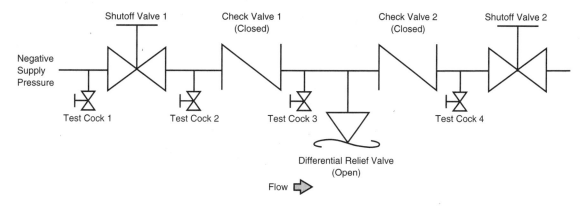

Figure 4-9 Backsiphonage: both check valves closed and the differential relief valve open

pressure less the relief-valve opening point (minimum of 2 psig [13.8 kPa gauge]), the relief valve will open and discharge water from the assembly to the atmosphere. This discharge from the relief valve ensures that the pressure after the first check is always lower than the inlet pressure.

In a backsiphonage condition, the inlet pressure will be reduced to a subatmospheric pressure. The pressure downstream from the first check which will cause the relief valve to open and discharge the water to atmosphere (Figure 4-9).

Application. The RP is an assembly that can prevent backflow from backpressure and/or backsiphonage. An RP is designed for both high- and low-hazard applications. An RP can be used for service protection or internal protection (see Figure 4-10).

Installation criteria. An RP must be installed in the orientation as it was approved by the approval agency recognized by the jurisdictional authority.

An RP must not be subjected to conditions that would exceed its maximum working water pressure and temperature rating. The increased pressure that can occur because of the creation of a closed system also must be evaluated, because excessive backpressure can damage the assembly or other plumbing components.

An RP should be sized hydraulically, taking into account both volume requirements and pressure loss through the assembly.

A pipeline should be thoroughly flushed before an RP is installed to ensure that no dirt or debris is delivered into the assembly, because dirt or debris may adversely affect the RP's working abilities.

An RP must not be installed in a pit or below grade where the relief valve could become submerged in water or where fumes could be present at the relief-valve discharge, because this installation may allow water or fumes to enter the assembly.

An RP shall be installed a minimum of 12 in. (305 mm) above the relief-valve discharge-port opening and the surrounding grade and floodplain (as regulated). The installation should not be installed where platforms, ladders, or lifts are required for access. If an assembly is installed higher than 5 ft (1.5 m) above grade, a permanent platform should be installed around the assembly to provide access for workers. (Check with local administrative authorities to confirm acceptability.)

An RP shall be installed where it can be easily tested and repaired as necessary. The assembly shall have adequate clearance around it to facilitate disassembly, repairs, testing, and other maintenance.

An RP may periodically discharge water from the relief valve. The effect of this discharge from the relief valve around the assembly must be evaluated. If the RP discharge is piped to a drain, an air-gap separation must be installed between the relief-valve

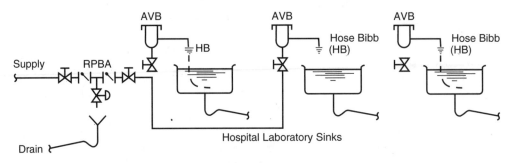

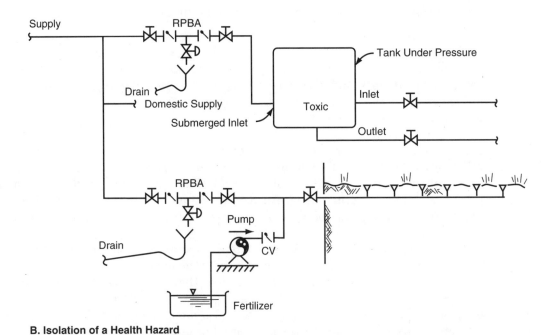

A. Area Isolation of a Health Hazard

B. Isolation of a Health Hazard

Figure 4-10 Typical reduced-pressure principle backflow-prevention applications

discharge opening and the drain line leading to the drain. Most RP manufacturers produce an air-gap drain fitting that attaches to the relief valve so that piping can be run from the RP to the drain. *Caution:* The air-gap drain is designed to carry occasional small amounts of discharge and/or spitting from the relief valve. The full discharge of an RP generally is beyond the capacity of the air-gap drain fitting.

If an RP must be subjected to environmental conditions that could freeze or heat the assembly beyond its working temperatures, some means of protection should be installed to provide the correct temperature environment in and around the assembly.

Reduced-Pressure Principle Detector Backflow-Prevention Assembly

Description. The reduced-pressure principle detector backflow-prevention assembly (RPDA) shall consist of a main-line RP with a bypass arrangement around the RP that shall contain a bypass water meter and a bypass RP. The RPDA shall be installed as an assembly as designed and constructed by the manufacturer. Although not recognized by this manual, other organizations provide for a bypass check in lieu of the bypass RP.

Function. The RPDA operates like an RP except the bypass is engineered to detect the first 3 gpm (11.4 L/min) of flow through the assembly. This low flow is registered by the water meter in the bypass and is used to show any unauthorized usage or leakage in the fire protection system.

Application. An RPDA is an assembly designed for fire protection systems for which a main-line meter is not used but the need to determine leaks or unwanted usage is desired. An RPDA can protect against backpressure and/or backsiphonage. An RPDA can be used to protect both high- and low-hazard installations. An RPDA can be used for service protection or internal protection.

Installation criteria. The criteria are the same as for the RP (see RP installation criteria).

Double Check Valve Backflow-Prevention Assembly

Description. The double check valve backflow-prevention assembly (DC) shall contain two internally loaded, independently operating, approved check valves; two resilient-seated shutoff valves; and four properly located test cocks (Figure 4-11). The DC shall be installed as an assembly as designed and constructed by the manufacturer. The assembly shall be approved by an approval agency acceptable to the local administrative authority.

Function. The check valves are designed to generate a loading able to hold a minimum of 1 psi (6.9 kPa) in the direction of flow of the check valve, with the outlet side of the check at atmospheric pressure. During normal operation, the check valves will open in response to demand for water at the outlet (Figure 4-12). When the demand for water ceases, the check valves will close.

In a backpressure condition, the increase of pressure on the outlet will cause the second check to close. If the second check does not seal properly, the first check will act as a backup to the second check (Figure 4-13).

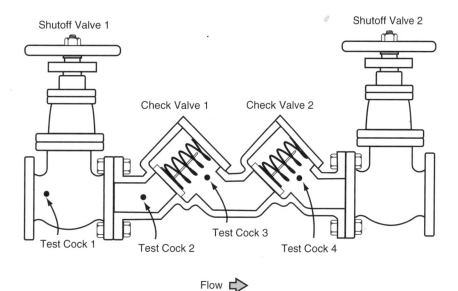

Figure 4-11   Double check valve assembly

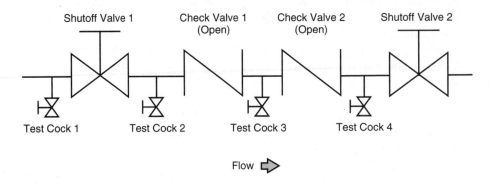

Figure 4-12   Check valves open, permitting flow

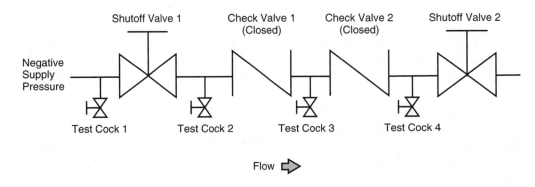

Figure 4-13   Negative supply pressure, check valves closed

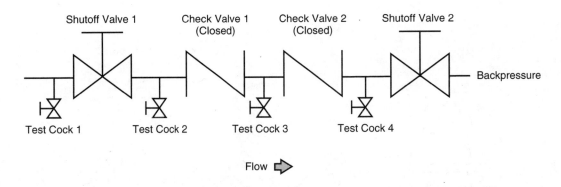

Figure 4-14   Backpressure: both check valves closed

In a backsiphonage condition, the inlet pressure will be reduced to a subatmospheric pressure. The greater pressure on the downstream side of the second check will cause the second check to close. If the second check does not seal off properly, the first check will act as a backup (Figure 4-14).

Application.   The DC is an assembly designed to prevent backflow from backpressure and/or backsiphonage. A DC can be used only for low-hazard applications. A DC can be used for service protection or internal protection (Figure 4-15).

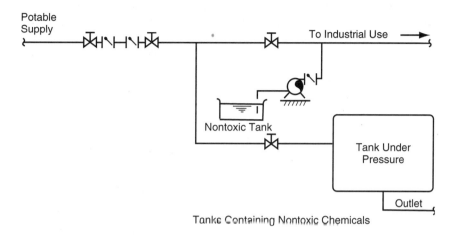

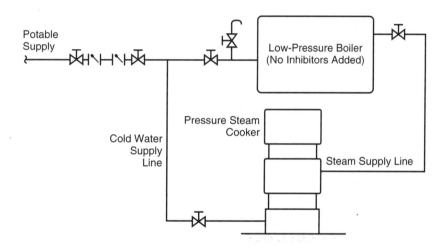

Figure 4-15 Typical double check valve assembly applications

Installation criteria. A DC must be installed in the orientation as it was approved by the approval agency recognized by the jurisdictional authority.

A DC must not be subjected to conditions that would exceed its maximum working water pressure and temperature rating. The increased pressure that can happen from the creation of a closed system also must be evaluated to prevent damage to the assembly or other plumbing-system components.

A DC should be sized hydraulically, taking into account both volume requirements and pressure loss through the assembly.

A DC should not be installed in a pit or below grade when possible. If the DC must be installed in a vault, adequate space for testing and maintenance must be provided. If the DC must be installed below grade, the test cocks shall be sealed or plugged so water or debris cannot collect in the test cock.

A pipeline should be thoroughly flushed before a DC is installed to ensure that no dirt or debris is delivered to the assembly, because dirt or debris may adversely affect the DC's working abilities.

A DC shall be installed a minimum of 12 in. (305 mm) above the surrounding grade and floodplain (as regulated). The installation should not be installed where platforms, ladders, or lifts are required for access. If an assembly must be installed higher than 5 ft (1.5 m) above grade, a permanent platform shall be installed around the assembly to provide access for workers. (Check with local administrative authorities.)

A DC shall be installed where it can be easily field-tested and repaired as necessary. The assembly shall have adequate clearance around it to facilitate testing, disassembly, and assembly of the DC.

If a DC must be subjected to environmental conditions that could freeze or heat the assembly beyond working temperatures, some means of protection should be installed to provide the correct temperature environment in and around the assembly.

Double Check Detector Backflow-Prevention Assembly

Description. The double check detector backflow-prevention assembly (DCDA) shall consist of a main-line DC with a bypass (detector) arrangement around the main-line DC that shall contain a bypass water meter and a bypass DC. The DCDA shall be installed as an assembly as designed and constructed by the manufacturer. Although not recognized by this manual, other organizations provide for a single bypass check in lieu of the bypass DC.

Function. The DCDA operates like a DC except the bypass is engineered to detect the first 3 gpm (11.4 L/min) of flow through the assembly. This low flow is registered by the water meter in the bypass and is used to show any unauthorized usage or leaks in the fire protection system.

Application. A DCDA is an assembly designed for fire protection systems for which a main-line meter is not used but the need to determine leaks or unwanted usage is desired and detected in the bypass. A DCDA can protect against backpressure and/or backsiphonage. A DCDA can be used only for low-hazard applications. A DCDA can be used for service protection or internal protection.

Installation criteria. The criteria are the same as for the DC (see DC installation criteria).

Pressure Vacuum-Breaker Assembly

Description. The pressure vacuum-breaker assembly (PVB) shall contain an independently operating, internally loaded check valve and an independently operating, loaded air-inlet valve located on the discharge side of the check valve (Figure 4-16). In addition, the PVB assembly shall have an inlet and outlet resilient-seated, fully ported shutoff valve and two properly located resilient-seated test cocks. The PVB shall be installed as an assembly as designed and constructed by the manufacturer. The PVB shall be approved by an approval agency acceptable to the local administrative authority.

Function. The check valve is designed to generate a loading able to hold a minimum of 1 psi (6.9 kPa) in the direction of flow with the outlet side of the check at atmospheric pressure. After water passes the check valve, it will cause the air-inlet poppet to close by overcoming the air-inlet loading, which is designed to be a minimum of 1 psi (6.9 kPa). During normal operation, the check valve will open in response to demand for water on the downside and the air inlet will remain closed. When the demand for water ceases, the check valve will close.

In a backsiphonage condition, the inlet pressure will be reduced to a subatmospheric pressure (Figure 4-17). The check valve will close because of the higher pressure on the downstream side of the check valve. When the pressure on the downstream side of the check valve falls to the air-inlet opening point (minimum of

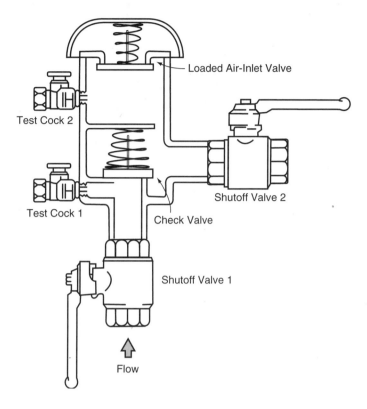

Figure 4-16　Pressure vacuum-breaker assembly

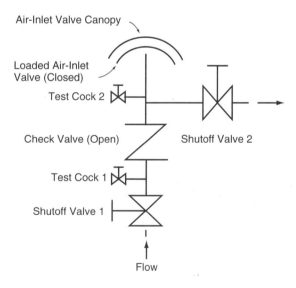

Figure 4-17　Pressure vacuum-breaker assembly, normal flow condition

1 psi [6.9 kPa]), the air inlet will open to ensure that any vacuum is broken. If the check valve does not seal properly, the area after the check valve will decrease in pressure, causing the air-inlet poppet to come off the air-inlet seat; this action will open and break any vacuum by allowing air to be siphoned into the plumbing system instead of the downstream water.

54 BACKFLOW PREVENTION AND CROSS-CONNECTION CONTROL

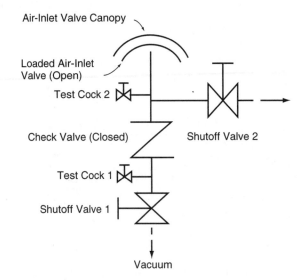

Figure 4-18   Pressure vacuum-breaker assembly, backsiphonage condition

Application. The PVB is an assembly designed to prevent backflow only from backsiphonage (Figure 4-18). The PVB can be used for both high- and low-hazard applications. The PVB can be used for internal protection. The PVB normally is not used for service protection because of its inability to protect against backpressure. However, a PVB may be used for service protection if the service is a dedicated, single-use service, such as an irrigation system.

Installation criteria. A PVB must be installed in the orientation as it was approved by the approval agency recognized by the jurisdictional authority.

A PVB must not be subjected to conditions that would exceed its maximum working water pressure and temperature rating. The increased pressure that can happen from the creation of a closed system also must be evaluated, because a PVB cannot be exposed to backpressure.

A PVB shall not be installed where it is subjected to backpressure.

A PVB should be sized hydraulically, taking into account both volume requirements and pressure loss through the assembly.

A pipeline should be thoroughly flushed before a PVB is installed to ensure that no dirt or debris is delivered into the assembly, because dirt or debris may affect the PVB's working abilities.

A PVB must not be installed in a pit or below grade where the air inlet could become submerged in water or where fumes could be present at the air inlet, because this installation may allow water or fumes to enter the assembly.

A PVB shall be installed a minimum of 12 in. (305 mm) above the highest point of use and any downstream piping supplied from the assembly. The installation should not be installed where platforms, ladders, or lifts are required for access. If an assembly must be installed higher than 5 ft (1.5 m) above grade, a permanent platform should be installed around the assembly to provide access for workers. (Check with local administrative authorities.)

A PVB shall be installed where it can be easily field-tested and repaired as necessary. The assembly shall have adequate clearance around it to facilitate disassembly, repairs, testing, and other maintenance.

A PVB may periodically discharge water from the air inlet. The effect of this discharge on the area around the assembly must be evaluated.

If a PVB must be subjected to environmental conditions that could freeze or heat the assembly beyond its working temperatures, some means of protection should be installed to provide the correct temperature environment in and around the assembly.

Spill-Resistant Vacuum Breaker

Description. The spill-resistant vacuum breaker (SVB) shall contain an internally loaded check valve and a loaded air-inlet valve located on the discharge side of the check valve. In addition, the SVB assembly shall have an inlet and outlet, resilient-seated, fully ported shutoff valve and a properly located resilient-seated test cock and vent valve. The SVB shall be installed as an assembly as designed and constructed by the manufacturer. The SVB shall be approved by an approval agency acceptable to the local administrative authority.

Function. The check valve is designed to generate a loading able to hold a minimum of 1 psi (6.9 kPa) in the direction of flow of the check valve with the outlet side of the check at atmospheric pressure. As water enters the assembly inlet, the pressure will first cause the air-inlet poppet to rise and seal against the air-inlet seat. After the air inlet is sealed, the check assembly will open, allowing water into the piping system. During normal operation, the check valve will open in response to demand for water on the outlet and the air inlet will remain closed. When the demand for water ceases, the check valve will close.

In a backsiphonage condition, the inlet pressure will be reduced to a subatmospheric pressure. The check valve will close and seal because of the higher pressure on the downstream side of the check valve. If the pressure in the SVB body is reduced by usage on the downstream side, the pressure will be relieved until the point the air-inlet poppet comes off the air-inlet seat and opens. If the check valve does not seal properly, the area after the check valve will decrease in pressure, causing the air inlet to open; this action will break any vacuum by allowing air to be siphoned into the plumbing system instead of the downstream water.

Application. The SVB is an assembly designed to prevent backflow only from backsiphonage. The SVB can be used for both high- and low-hazard applications. Normally, the SVB is authorized only for internal protection, for both indoor and outdoor application.

Installation criteria. An SVB must be installed in the orientation as it was approved by the approval agency recognized by the authority having jurisdiction.

An SVB must not be subjected to conditions that would exceed its maximum working water pressure and temperature rating. The increased pressure that can happen from the creation of a closed system also must be evaluated, because an SVB cannot be exposed to backpressure.

An SVB shall not be installed where it is subjected to backpressure.

An SVB should be sized hydraulically, taking into account both volume requirements and pressure loss through the assembly.

A pipeline should be thoroughly flushed before an SVB is installed to ensure that no dirt or debris is delivered into the assembly, because dirt or debris may adversely affect the SVB's working abilities.

An SVB must not be installed in a pit or below grade where the air inlet could become submerged in water or where fumes could be present at the air inlet, because this installation may allow water or fumes to be siphoned into the assembly.

An SVB shall be installed a minimum of 12 in. (305 mm) above the highest point of use and any downstream piping coming from the assembly. Installation heights may be varied. Check with local administrative authorities for optional acceptance criteria.

The installation should not be installed where platforms, ladders, or lifts are required for access. If an assembly must be installed higher than 5 ft (1.5 m) above grade, a permanent platform should be installed around the assembly to provide worker access. (Check with local administrative authorities for variances.)

An SVB shall be installed where it can be easily tested and repaired as necessary. The assembly shall have adequate clearance around it to facilitate disassembly, repairs, testing, and other maintenance.

BACKFLOW DEVICES

Backflow devices are not to be substituted for applications that require backflow assemblies. They usually do not include shutoff valves or test cocks, and they usually cannot be tested or repaired in-line. Many devices have restrictive head-loss and flow restrictions. These devices are used for internal protection and usually come only in smaller sizes (2 in. [51 mm] and smaller). The application of these devices usually comes under the jurisdiction of the plumbing code, because they usually are located only in private plumbing systems. A water purveyor that has concerns about the use of some devices should consult with local plumbing code officials concerning whether, what types, and where backflow devices can be used.

Atmospheric Vacuum Breaker

Description. An atmospheric vacuum breaker (AVB) shall contain an air-inlet valve and a check seat. The device shall be approved by an approval agency acceptable to the local administrative authority.

Function. Water will enter the inlet of the AVB and cause the air-inlet poppet to seal against the air-inlet seat (Figure 4-19). After the poppet is sealed, water will flow through the AVB into the piping system. In a backsiphonage situation, the inlet pressure will be reduced to a subatmospheric pressure, causing the poppet to fall off the air-inlet seat and rest on the check seat. Air will enter the air-inlet port(s) to break any vacuum.

Application. An AVB shall be installed to prevent backflow from backsiphonage only. An AVB can protect both high- and low-hazard applications. An AVB is used for internal protection, for both indoor and outdoor application. A wide variety of AVBs are produced for specific installations, such as lab faucets and for many types of equipment that use built-in AVBs.

Installation criteria. An AVB must be installed in the orientation as it was approved by the approval agency recognized by the authority having jurisdiction.

An AVB must not be subjected to conditions that would exceed its maximum working water pressure and temperature rating.

An AVB shall not be installed where it could be subjected to backpressure.

A pipeline should be thoroughly flushed before an AVB is installed to ensure that no dirt or debris is delivered into the device, because dirt or debris may adversely affect the AVB's working abilities.

An AVB must not be installed in a pit or below grade where the air inlet could become submerged in water or where fumes could be present at the air inlet, because this installation may allow water or fumes to enter the device.

An AVB shall be installed a minimum of 6 in. (152.5 mm) above the highest point of use and any downstream piping coming from the device. Installation heights may vary. Check with the local administrative authority for optional acceptance criteria.

An AVB shall not be subjected to continual use and shall not be pressurized more than 12 hours in a 24-hour period.

No control valves shall be on the downstream side of the AVB device.

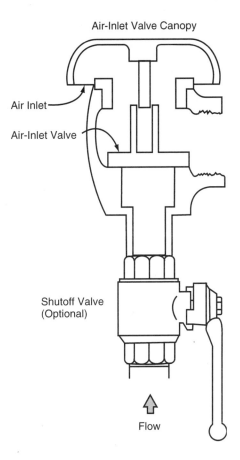

Figure 4-19 Atmospheric vacuum breaker

Dual Check

Description. A dual check shall contain two internally loaded, independently operating check valves. A dual check shall be a device approved by an approval agency acceptable to the local administrative authority.

Function. A dual check contains two loaded checks (Figure 4-20). In a backpressure condition, the increase in pressure will cause the checks to close. If the second check is not working, the first check can act as a backup to stop the backpressure from going through the device. In a backsiphonage condition, a subatmospheric condition is present at the inlet, and the loading of the checks will cause the checks to close.

Application. A dual check can be used to stop backflow from backpressure and/or backsiphonage. A dual check should be used only for low-hazard internal-protection applications. A dual check does not meet the requirements of a backflow-prevention assembly.

Installation criteria. A dual check must be installed in the orientation as it was approved by the approval agency recognized by the jurisdictional authority.

A dual check must not be subjected to conditions that would exceed its maximum working water pressure and temperature rating. The increased pressure that can happen from the creation of a closed system also must be evaluated, because excessive pressure can damage the device or other plumbing components.

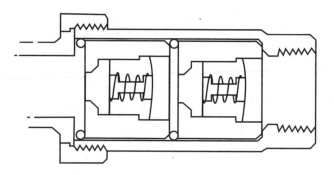

Figure 4-20 Dual check device

A dual check should be sized hydraulically, taking into account both volume requirements and pressure loss through the device.

A pipeline should be thoroughly flushed before a dual check is installed to ensure that no dirt or debris is delivered into the device, because dirt or debris may adversely affect the dual check's working abilities.

A dual check shall be installed where it can be inspected or replaced as necessary.

Dual Check With Atmospheric Vent

Description. The device shall contain two internally loaded check valves and a vent valve; the vent valve, located between the two check valves, shall open when subjected to backpressure. A dual check with atmospheric vent shall be a device approved by an approval agency acceptable to the local administrative authority.

Function. The device shall have an inlet check valve that opens as flow begins. After the water has passed the first check, it shall cause the vent valve to close, allowing water to travel past the second check valve. In a backpressure condition, the increased pressure at the outlet will cause the second check valve to close. If the check does not close, the increased pressure will travel across the second check and cause the vent valve to open as it is subjected to backpressure. In a backsiphonage condition, the inlet pressure will be reduced to a subatmospheric pressure and cause the check valves to close.

Application. A dual check with atmospheric vent can prevent backflow from backpressure and/or backsiphonage. The device should be used only for low-hazard applications. The device should be used only for internal protection. A dual check with atmospheric vent does not meet the requirements of a backflow-prevention assembly.

Installation criteria. A dual check with atmospheric vent must be installed in the orientation as it was approved by the approval agency recognized by the jurisdictional authority.

A dual check with atmospheric vent must not be subjected to conditions that would exceed its maximum working water pressure and temperature rating. The increased pressure that can happen from the creation of a closed system also must be evaluated. A dual check with atmospheric vent should be sized hydraulically, taking into account both volume requirements and pressure loss through the device.

A pipeline should be thoroughly flushed before a dual check with atmospheric vent is installed to ensure that no dirt or debris is delivered into the device, because dirt or debris may adversely affect its working abilities.

A dual check with atmospheric vent shall be installed where it can be inspected or replaced as necessary.

A dual check with atmospheric vent may discharge water from its vent. Care should be taken to ensure that any discharge will not harm the surrounding area.

TESTING

Continuous Water Service

When backflow-prevention assemblies are installed, they must be field-tested and maintained. This means that the water must be shut off temporarily for these important functions. If the assembly is feeding a critical service facility where water cannot be interrupted for a period of time, it is strongly suggested that more than one assembly be installed in a parallel installation. In this way, one assembly can be field-tested or repaired while the facility receives water that is delivered by the other assembly. Two or more parallel assemblies shall be the same type of assembly, as the degree of hazard shall dictate. When parallel assemblies are used, the piping system should be sized hydraulically, taking into account both volume requirements and pressure loss through the assemblies.

Testing Awareness

Backflow-prevention assemblies are installed in locations where a reliable means of backflow protection has been determined to be necessary. To ensure the continued protection of an identified actual or potential cross-connection, the properly installed backflow-prevention assemblies must be field-tested at least annually to ensure that they continue to prevent against backflow. If, in the testing process, the assembly is found to be working at less than its minimum criteria, it shall be repaired and returned to its full ability to prevent backflow. A person shall be allowed to field-test backflow-prevention assemblies only after successfully completing a training course and passing a certification course for backflow-prevention assembly testers. Some jurisdictions may require additional business and/or performance licensing to perform the duties of a backflow-prevention assembly tester.

Backflow-prevention assemblies may need to be field-tested more than annually. In addition to annual field testing, backflow-prevention assemblies are field-tested

- immediately following initial installation,
- whenever an assembly is relocated,
- whenever supply piping is altered,
- whenever an installed assembly is newly discovered and previous testing records are not available,
- whenever an assembly is taken apart for repair,
- whenever the administrative authority requires more frequent testing to ensure continued protection, and
- whenever the assembly has been taken out of service and is returned to service.

Testing Procedures of Assemblies

The purpose of the field-test procedure is to collect accurate data about the workings of a backflow-prevention assembly. The data collected from the field test must be compared to minimum acceptable standards to ensure that the backflow-prevention assembly is performing properly and can be relied on to prevent backflow in the field. If the data from the field test shows that the backflow-prevention assembly does not meet the minimum criteria, the assembly shall be repaired or replaced.

Because of the many variations in specific, detailed field-test procedures and because various regulatory authorities accept different field-test procedures, this manual will not specify any specific field-test procedure as accepted in lieu of other locally approved procedures. (See appendix A for various sample field-test procedures, which are provided for information only.) The local administrative authority shall evaluate available field-test procedures and specify a single field-test procedure for each type of assembly that will be used by testers in their jurisdiction.

The administrative authority must be assured that testers in its jurisdiction can properly perform field-test procedures on backflow-prevention assemblies. To provide this assurance, each tester must successfully complete a recognized examination process. This examination shall include testing about the theory of backflow prevention and a hands-on, practical examination of field-test procedures on all types of backflow-prevention assemblies in both properly working and malfunctioning conditions. The successful completion of this examination should lead to a certification that allows the tester to perform field tests of backflow-prevention assemblies and to submit field-test data to the administrative authority. This certification shall be valid for a specific length of time. When it expires, the tester shall be required to obtain recertification by successfully completing another similar examination process that combines testing about theory of backflow prevention and a hands-on, practical examination of field-test procedures on all types of backflow-prevention assemblies in both properly working and malfunctioning conditions.

Following is a list of agencies and publications that provide field-test procedures for backflow-prevention assemblies:

- American Backflow Prevention Association (ABPA) Tester Certification Program
- American Society of Sanitary Engineering (ASSE) Professional Qualifications Standards, Backflow Prevention Assemblies—Series 5000
- Canadian Standards Association (CSA)
- USC Foundation for Cross-Connection Control and Hydraulic Research, *Manual of Cross-Connection Control*, 9th edition
- New England Water Works Section of AWWA, *Backflow Device Testing Procedures*
- USEPA, *Cross-Connection Control Manual*, 1989
- University of Florida TREEO Center, *Backflow Prevention: Theory & Practice*

After the certified tester has properly performed the field-test procedure on a backflow-prevention assembly, the data collected shall be recorded on a field-test report form. The data on this field-test report form shall properly represent the workings of the assembly at the time it was tested by the certified tester. Copies of the field-test report form should be sent to all appropriate interested parties including, as a minimum, the water user and the local administrative authority. The tester shall retain copies of the test data. Records should be maintained for at least 3 to 5 years.

Backflow-Prevention Assembly Field-Test Equipment

To properly collect accurate data in the field about the operation of a backflow-prevention assembly, a certified tester must properly adhere to a proper field-test procedure. The last variable in the data-collection process is the field-test equipment used to field-test backflow-prevention assemblies. This field-test equipment must be capable of providing accurate data. There are many types of commercially available field-test equipment. The water purveyor must verify that the field-test equipment used can perform the field-test procedure that is accepted in its jurisdiction and that the accuracy of the equipment is maintained. To ensure the ability of the test equipment to provide accurate data, the accuracy of the field-test equipment shall be verified at least annually. If the accuracy of the field-test equipment is not within accepted standards recognized by the local administrative authority, the field-test equipment shall be calibrated and brought into acceptable accuracy tolerances.

Maintenance of assemblies. When a certified tester performs an accurate field test on an approved backflow-prevention assembly with accurate field-test equipment, the data shall properly represent the working abilities of the backflow preventer. When this data indicates that the assembly is not performing to the minimum accepted standards as established by the field-test procedure, the results shall be noted on the field-test report, and the assembly shall be repaired or replaced to ensure that the assembly will continue to prevent backflow. The purpose of a repair procedure is to return the assembly to a condition in which it will reliably protect against backflow. All backflow-prevention assemblies are designed to be repaired. Manufacturers provide repair parts.

The first step in the repair process shall be obtaining the necessary repair information. Most manufacturers provide factory repair information about the repair processes for their assemblies. This information should be reviewed. The next recommended step is to obtain original factory repair parts. The repair process will involve the disassembly of the backflow preventer and the proper placement of the new repair parts. Care must be taken to safely perform the repair. After the repair is performed, the backflow-prevention assembly shall be field-tested by a certified tester to ensure that the repair procedure has restored the assembly to proper working condition.

It may not be possible to repair some backflow-prevention assemblies. This usually involves assemblies that do not have replaceable parts, such as check seats, that can be cast into the body of some of the older assemblies. The original manufacturer may no long produce repair parts for some older assemblies. In these cases, the assembly may need to be replaced. When replacing an existing assembly with a new one, it may be necessary to survey the application to ensure that the proper type of backflow-prevention assembly is installed for the application.

Tester Responsibilities

Accurate data about the workings of properly installed (and approved) backflow-prevention assemblies is important to the administration of an effective cross-connection control program. The tester is responsible for the accurate generation of data, a correct assessment of the workings of each assembly tested, and proper dissemination of the data to all necessary parties. Certified testers shall adhere to the field-test procedures established in the certification process. Local administrative authorities shall review the data submitted by the certified testers to review the results and interpret the data to ensure the tester is performing the field test properly. Local administrative authorities may elect to establish local criteria in addition to the procedures established by the accepted certification program. If local administrative authorities choose to establish additional criteria, these criteria must be communicated to testers. A periodic meeting or other contact with testers can help to clarify the requirements.

This page intentionally blank.

AWWA MANUAL M14

Chapter 5

Typical Hazards

As discussed in chapter 2, the water purveyor has the responsibility to protect the water supplied to the customer from contamination caused by cross-connection hazards posed by both the customer's plumbing system and the water purveyor's collection, treatment, and distribution system. The first part of this chapter discusses the installation, by the water purveyor or the customer, of approved backflow preventers on the water service to eliminate backflow from the customer's internal water system (i.e., plumbing) to the public water system. The second part of this chapter discusses the installation of approved backflow prevention within the water purveyor's distribution system and other public water system facilities.

The responsibilities to evaluate internal hazards and to require the separation of potable water piping from piping, equipment, fixtures, operations, or other water uses that might pollute or contaminate the potable water (internal protection) rest with the health agency and/or the regulatory agency that has jurisdiction. This chapter's discussion of the typical hazards found within the customer's premises is provided only for the water purveyor's overall assessment of the customer's plumbing system. It is not intended to provide guidance regarding internal protection for compliance with the plumbing code or with other regulations pertaining to the customer's property.

Many chemical and bacteriological substances, foreign materials, and sources of unknown water quality create actual or potential hazards to the public water system when introduced into the consumer's plumbing system. However, these hazards are not all-inclusive and should not preclude the evaluation of every water service for potential backflow hazards. Unless stipulated otherwise, the following backflow-prevention recommendations are to be applied to water service connections.

RECOMMENDED PROTECTION FOR SPECIFIC CATEGORIES OF CUSTOMERS

Auxiliary Water Systems

An approved backflow-prevention assembly shall be installed at the water purveyor's service connection to any premises where there is an auxiliary water supply or system, even though there is no connection between the auxiliary water supply and the public potable water system.

The term *auxiliary water supply* (defined in the glossary) is commonly used to describe water supplies or sources that are not under the water purveyor's control or direct supervision. Typical of such water supplies are natural waters derived from wells, springs, streams, rivers, lakes, harbors, bays, and oceans. Also considered auxiliary water supplies (i.e., other than the public potable water supply that is under the water purveyor's control) are used waters that have passed beyond the water purveyor's control at the point of delivery and that may be stored, transmitted, or used in a way that may contaminate them. Finally, public potable water supplies furnished by another water purveyor may not be under good sanitary control or may be otherwise unacceptable to the water purveyor; these would also be considered auxiliary supplies.

Typical used water supplies include

- water in industrialized water systems;
- water in reservoirs or tanks used for fire-fighting purposes;
- irrigation reservoirs;
- swimming pools, fish ponds, and mirror pools;
- memorial and decorative fountains and cascades;
- cooling towers;
- baptismal, quenching, washing, rinsing, and dipping tanks;
- reclaimed water and recycled water;
- gray water and spray aerobic water.

All of these supplies, including a public potable water supply over which the water purveyor does not exercise sanitary control, are potential hazards to the public water system. These waters may become contaminated because of industrial processes; through contact with the human body, dust, vermin, birds, etc.; or by means of chemicals or organic compounds that may be introduced into tanks, lines, or systems for control of scale, corrosion, algae, bacteria, or odor or for similar treatment.

In evaluating the hazard that an auxiliary water supply poses to the water purveyor's system, it is not necessary to determine that the auxiliary water sources are developed and connected to the potable water system through cross-connections. It is only necessary to determine that the water or fluids are available to the premises and are of a quantity sufficient to make it desirable and feasible for the customer to develop and use the supply.

For administrative purposes, the water purveyor may subdivide auxiliary water supplies into three general classifications: an approved public potable water supply over which the water purveyor does not have sanitary control; any private water supply, other than the water purveyor's approved public potable water supply, that is on or available to the premises; and used waters and industrial fluids, such as waters in reservoirs, cooling towers, recirculation systems, industrial-fluid systems, gray water, reclaimed water, and the like.

Protection recommended. For connections to potable water systems:

- An air-gap separation or a reduced-pressure principle backflow-prevention assembly is recommended at the service connection when the auxiliary water supply is or may be contaminated to a degree that would constitute a high hazard.
- A double check valve assembly is recommended at the service connection when the auxiliary water supply is being operated under a public health permit but is not acceptable to the water purveyor as a source.

No backflow protection at the service connection is required if the permitted auxiliary water system has a properly conducted sanitary control program in force and if the auxiliary water supply is acceptable to the water purveyor as a source. However, it is recommended that the water purveyor consider installing a double check valve assembly to eliminate the return of any water that is not completely under the water purveyor's control.

For private water supplies:

- An air-gap separation or a reduced-pressure principle backflow-prevention assembly is recommended at each service connection when the auxiliary water supply is or may be contaminated to a degree that it would constitute a high hazard.
- A double check valve assembly is recommended at all other service connections.

For used waters and industrial fluids:

- An air-gap separation or a reduced-pressure principle backflow-prevention assembly is recommended where there is a high hazard.
- A double check valve assembly is recommended at all other service connections.

Beverage-Bottling Plants and Breweries

An approved backflow-prevention assembly shall be installed at the service connection to any premises where a beverage-bottling plant is operated or maintained and water is used for industrial purposes.

The hazards normally found in plants of this type include cross-connections between the potable water system and steam-connected facilities, such as pressure cookers, autoclaves, retorts; washers, cookers, tanks, lines, flumes, and other equipment used for storing, washing, cleaning, blanching, cooking, flushing, or fluming, or for transmission of foods, fertilizers, or wastes; can- and bottle-washing machines; lines where caustics, acids, detergents, and other compounds are used in cleaning, sterilizing, and flushing; reservoirs, cooling towers, and circulating systems that may be heavily contaminated with vermin, algae, bacterial slimes, or toxic water-treatment compounds, such as pentachlorophenol, copper sulfate, chromates, metallic glucosides, compounds of mercury, quaternary ammonium compounds, etc.; steam-generating facilities and lines that may be contaminated with boiler compounds, such as those chemicals listed above (NOTE: A very particular hazard is the possibility that steam will get back into the potable system, causing system damage or a health hazard.); industrial-fluid systems and lines containing cutting and hydraulic fluids, coolants, hydrocarbon products, glycerin, paraffin, caustic and acid solutions, etc.; water-cooled equipment that may be sewer-connected, such as compressors, heat exchangers, air-conditioning equipment; fire-fighting systems, including storage reservoirs that may be treated for prevention of scale formation, corrosion, algae, slime growths, etc.; fire systems that may be subject to contamination with antifreeze solutions, liquid foam concentrates, or other chemicals or chemical compounds used in fighting fire; fire systems that are subject to contamination with auxiliary or used water supplies; or industrial fluids.

Protection recommended. An air-gap separation or a reduced-pressure principle backflow-prevention assembly is recommended where there is a high hazard. A double check valve assembly should be used for all other connections.

Canneries, Packing Houses, Food Service Facilities, Restaurants, and Reduction Plants

An approved backflow-prevention assembly shall be installed at the service connection to any premises where fruit, vegetable, or animal matter is canned, concentrated, processed, or served.

The hazards normally found in facilities of this type include cross-connections between the potable water system and steam-connected facilities, such as pressure cookers, autoclaves, retorts; post-mix beverage-dispensing machines, washers, cookers, tanks, lines, flumes, and other equipment used for storing, washing, cleaning, bleaching, cooking, flushing, or fluming, or for transmission of food, fertilizers, or wastes; reservoirs, cooling towers, and circulating systems that may be heavily contaminated with vermin, algae, bacterial slimes, or toxic water-treatment compounds, such as pentachlorophenol, copper sulfite, chromates, metallic glucosides, compounds of mercury, quaternary ammonium compounds; steam-generating facilities and lines that may be contaminated with boiler compounds, such as the chemicals listed above (NOTE: A very particular hazard is the possibility that steam will get back into the potable system, causing a health hazard or system damage.); industrial-fluid systems and lines containing cutting and hydraulic fluids, coolants, hydrocarbon products, glycerin, paraffin, caustic and acid solutions, etc.; fire-fighting systems, including storage reservoirs that may be treated for prevention of scale formation, corrosion, algae, slime growths, etc.; fire systems that may be subject to contamination with antifreeze solutions, liquid foam concentrates, or other chemicals or chemical compounds used in fighting fire; fire systems that are subject to contamination with auxiliary or used water supplies or industrial fluids; water-cooled equipment that may be sewer-connected, such as compressors, heat exchangers, air-conditioning equipment; and tanks, can- and bottle-washing machines, and lines where caustics, acids, detergents, and other compounds are used in cleaning, sterilizing, and flushing.

Protection recommended. An air-gap separation or a reduced-pressure principle backflow-prevention assembly is recommended.

Chemical Plants and Other Facilities: Cleaning, Manufacturing, Processing, Compounding, Servicing, Treatment, or Washing

This category covers service connections to any premises where there is a facility where water is used in the industrial processes of manufacturing, storing, compounding, cleaning, or processing chemicals. This category also includes facilities where chemicals are used as an additive to the water supply or are used in processing or in servicing products.

This is a very broad category. The water purveyor must give careful consideration to the operational processes involved in these facilities. Water is required for manufacturing purposes in most chemical plants; it is used to purge lines and to clean vats and tanks, as well as for process water. Cross-connections in such plants may be numerous because of the intricate piping. The severity of the cross-connection hazards varies with the toxicity of chemicals used.

The hazards normally found in facilities of this type include cross-connections between the potable water system and formulating tanks and vats, decanter units, extractor/precipitators, and other processing units that may be heavily contaminated with highly toxic end products or toxic waste by-products, including such contaminants as organophosphate derivatives, organo-nitrogen compounds, chlorinated-aimes, chlorinated-dibenzonfurans; reservoirs, cooling towers, and circulating systems that may be heavily contaminated with vermin, algae, bacterial slimes, or toxic water-treatment compounds, such as chromates, pentachlorophenol, copper sulfate, metallic glucosides, compounds of mercury, quaternary ammonium compounds; steam-generating facilities and lines that may be contaminated with boiler compounds, such as those chemicals listed above (NOTE: A very particular hazard is the possibility that steam will get back into the potable system, causing system damage or a health hazard.); industrial-fluid systems and lines containing cutting and hydraulic fluids, coolants, hydrocarbon products, glycerin, paraffin, caustic and acid solutions, etc.; fire-fighting systems, including storage reservoirs that may be treated for prevention of scale formation, corrosion, algae, slime growths, etc.; fire systems that may be subject to contamination with antifreeze solutions, liquid foam concentrate or other chemicals or chemical compounds used in fighting fire; fire systems that are subject to contamination with auxiliary or used water supplies or industrial fluids; water-cooled equipment that may be sewer-connected, such as compressors, heat exchangers, air-conditioning equipment; hydraulically operated equipment for which the water purveyor water pressure is directly used and may be subject to backpressure; equipment under hydraulic tests, such as tanks, lines, valves, and fittings; pumps, pressure cylinders, or other hydraulic facilities that may be used to provide pressure for testing purposes (NOTE: In such cases, air, gas, or hydraulic fluids may be forced back into the public system.); pressure cookers, autoclaves, retorts, and other similar steam-connected facilities; and washers, pressure washers, cookers, tanks, flumes, and other equipment used for storing, washing, cleaning, blanching, cooking, flushing, or fluming, or for transmission of foods, fertilizers, or wastes.

Protection recommended. An air-gap separation or a reduced-pressure principle backflow-prevention assembly is recommended.

Cooling Systems: Open or Closed

Cooling systems—including cooling towers—usually require some treatment of the water for algae, slime, or corrosion control.

Toxic chemicals frequently used for this purpose may include quaternary ammonium compounds, pentachlorophenol, mercury, or chromium; also used are chemicals that are toxic in higher concentrations, including chlorine, bromine, copper, permanganate, glucosides, and ozone.

Protection recommended. An air-gap separation or a reduced-pressure principle backflow-prevention assembly is recommended.

Dairies and Cold-Storage Plants

An approved backflow-prevention assembly shall be installed on the service connection to any premises on which a dairy, creamery, ice cream plant, or cold-storage plant is operated or maintained, or where commercial ice manufacturing equipment is used, provided such a plant has on the premises an industrial-fluid system, sewage handling facilities, or other similar source of contamination, that, if cross connected, would create a hazard to the public system.

The hazards normally found in plants of this type include cross-connections between the potable water system and reservoirs, cooling towers, and circulation systems that may be heavily contaminated with vermin, algae, bacterial slimes, or toxic water-treatment compounds, such as pentachlorophenol, copper sulfate, chromates, metallic glucosides, compounds of mercury, quaternary ammonium compounds; steam-generating facilities and lines that may be contaminated with boiler compounds, such as those chemicals listed above (NOTE: A very particular hazard is the possibility that steam will get back into the potable system, causing system damage or a health hazard.); water-cooled equipment that may be sewer-connected, such as compressors, heat exchangers, air-conditioning equipment; and tanks, can- and bottle-washing machines, pressure washers, and lines where caustics, acids, detergents, and other compounds are used in cleaning, sterilizing, and flushing.

Protection recommended. An air-gap separation or a reduced-pressure principle backflow-prevention assembly is recommended.

Dry-Pipe Nonpressurized Fire-Suppression Systems (Deluge)

Dry-pipe nonpressurized systems are open to the atmosphere; sprinkler or other outlets are open, ready to flow water at all times. Generally, dry-pipe nonpressurized fire-suppression systems that are directly connected to a public potable water main line do not present a health hazard to the public water system unless chemicals are added to the water as it enters the system.

Protection recommended. Generally, no backflow protection is required if there is not an addition of chemicals. Where chemicals are added or are likely to be added, a reduced-pressure principle backflow-prevention assembly, a reduced-pressure principle detector backflow-prevention assembly, or an air gap is recommended.

NOTE: As with existing wet-pipe systems, the backflow protection recommended for dry-pipe systems applies to new systems. Generally, it does not apply to existing systems that have some form of acceptable directional flow-control protection in place until those systems are substantially altered. A single check valve or alarm check valve is a directional control device and not a backflow preventer.

Dry-Pipe Pressurized and Preaction Fire-Suppression Systems

Dry-pipe pressurized systems typically are charged with air or nitrogen. Sprinklers or other outlets are closed until there is sufficient heat to open a sprinkler; then air is released and followed by water, which extinguishes the fire. Preaction systems contain air that may or may not be under pressure. They use automatic sprinklers and have a supplemental fire-detection system installed in the same areas as the sprinklers. These systems present a low hazard.

Protection recommended. A reduced-pressure principle backflow-prevention assembly is recommended where there is a high hazard (e.g., risk of chemical addition). A double check valve assembly should be used on all other systems.

NOTE: As with existing wet-pipe systems, the backflow protection recommended for dry-pipe systems applies to new systems and does not generally apply to existing systems that have some acceptable form of directional flow control in place until the systems are substantially altered.

Dye Plants

Most solutions used in dyeing are highly toxic. The toxicity depends on the chemicals used and their concentrations. The following types of chemical groups of dyes are generally used: vat dye, mordant dye, chrome dye, nitro dye, metallized dye, Thiazol dye, and anaine dye.

Protection recommended. An air-gap separation or a reduced-pressure principle backflow-prevention assembly is recommended.

Film Laboratories: Photo and X-ray

An approved backflow-prevention assembly shall be installed on each service connection to any premises where a film laboratory, film processing, or film manufacturing plant is operated or maintained. This does not include small, personal, noncommercial darkroom facilities.

The hazards normally found in plants of this type include cross-connections between the customers' water system and tanks, automatic film-processing machines, or other equipment used in processing that may be contaminated with chemicals, such as acetic acid, potassium ferricyanide, or one of the many types of the aromatic series of organic chemicals; or water-cooled equipment that may be sewer-connected, such as compressors, heat exchangers, and air-conditioning equipment.

Protection recommended. An air-gap separation or a reduced-pressure principle backflow-prevention assembly is recommended.

Fire Hydrants

Fire hydrants are installed primarily to provide a water supply for fire-fighting purposes. However, fire hydrants also provide access for contaminants to enter the water distribution system. To ensure public health and safety, fire hydrants supplied from potable water systems should be monitored regularly and maintained as required by the water purveyor and the fire authority. The water purveyor must also consider fire-fighting equipment and the use of chemicals with that equipment to ensure the potable water system is not contaminated.

Fire hydrants are used for purposes other than fire fighting, e.g., for construction water, dust control, water hauling, jumper connections for superchlorination of mains, pressure testing, and temporary service. These uses present the potential for many different types of cross-connection hazards to occur.

Protection recommended. Temporary service connections from a hydrant should be equipped with backflow protection commensurate to the degree of hazard presented. If there is any question regarding the degree of hazard, a reduced-pressure principle backflow-prevention assembly is recommended. This will ensure that the potable water supply is protected from unknown materials, chemicals, and other non-potable substances that are contained in, or that have been in contact with, hoses, pipes, tanks, etc.

NOTE: On mobile tanks, an air gap remains a good additional safeguard. However, because it does not ensure the integrity of the materials supplying the tank, it should be accepted as a sole means of backflow protection only if the water purveyor is certain that the piping or other conduit to the air gap will remain in a potable state at all times.

Fire Sprinkler Systems, Commercial

Previous editions of this manual classified fire-suppression systems into six classifications based on the complexity of the individual system. For purposes of this manual, to clarify and simplify the selection of appropriate backflow protection, the classifications have been eliminated.

The water purveyor must be mindful of state or provincial regulations pertaining to fire sprinkler systems. These regulations may limit the water purveyor's options for requiring backflow preventers on fire service lines. The water purveyor also must understand that installing a backflow preventer on an existing fire-suppression system may have a significant adverse effect on the hydraulic performance of the system. This is especially relevant when the original design may not have included such equipment. However, the cross-connection risks posed by new and existing fire-suppression systems are equal, and thus, there is no justification for different levels of public health protection.

Two approaches may be used to assess the water purveyor's hazard from its customers' fire-suppression systems. The first approach is to consider all types of fire-suppression system (e.g., wet pipe, dry pipe) hazards that require either a reduced-pressure principle backflow-prevention assembly or a double check valve assembly. The second approach is to make a detailed assessment of each type of fire-suppression system.

Where the first approach is followed, the water purveyor recognizes that any fire-suppression system may be operated in a manner other than the manner for which it was originally designed. For example, a dry-pipe system may inappropriately be operated as a wet system during most of the year, then be charged with air for the months of freezing weather. Or, a maintenance contractor could inappropriately add antifreeze to a wet system that was designed to operate without added chemicals. (NOTE: Such antifreezes may be of various grades, including food grade. See National Fire Protection Association Manual 13, *Installation of Sprinkler Systems* [2002].)

Protection recommended. For the first approach, a reduced-pressure principle backflow-prevention assembly is recommended where there is a high hazard (e.g., risk of chemical addition). A double check valve assembly should be used on all other systems.

Where the second approach is followed, subsequent sections of this chapter provide guidance for the following types of fire-suppression systems:

- New wet-pipe fire sprinkler systems
- Existing wet-pipe fire sprinkler systems
- Dry-pipe nonpressurized fire-suppression systems
- Dry-pipe pressurized fire-suppression systems
- Other fire-suppression systems

Hospitals, Laboratories, Medical Offices and Facilities, Medical Research Centers, Sanitariums, Morgues, Mortuaries, Autopsy Facilities, and Other Human or Animal Clinics

The hazards normally found in facilities of this type include cross-connections between the potable water system and contaminated or sewer-connected equipment, such as bedpan washers, flush-valve toilets and urinals, autoclaves, specimen tanks, development tanks, sterilizers, pipette-tube washers, cuspidors, aspirators, and autopsy and mortuary equipment. It has been found that in this type of facility, little or no attention is given to the maintenance of air-gap separations or vacuum breakers. It is customary to bridge an air-gap separation by means of a hose. Also, in multistoried buildings, the supply line to the toilets, urinals, lavatories, laboratory sinks, etc., on the lower floors may be taken off the suction side of the hose pump and, as a result, sewage or other contaminated substances may be drawn into the hose supply line; water-cooled equipment that may be sewer-connected, such as compressors, heat exchangers, and air-conditioning equipment; reservoirs, cooling towers, and circulating systems that may be heavily contaminated with vermin, algae, bacterial slimes, or toxic water-treatment compounds, such as pentachlorophenol, copper sulfate, chromate, metallic glucosides, compounds of mercury, and quaternary ammonium compounds; and steam-generating facilities and lines that may be contaminated with boiler compounds, such as those chemicals listed above (NOTE: A very particular hazard is the possibility that steam will get back into the potable system, causing system damage or a health hazard.).

Protection recommended. An air-gap separation or a reduced-pressure principle backflow-prevention assembly shall be installed on the service connection to any hospital, mortuary, morgue, autopsy facility, other facilities as listed, or any multistoried medical building or clinic.

Irrigation Systems

Irrigation systems include but are not limited to agricultural, residential, and commercial applications; they may be connected directly to the public water system or internally to the private plumbing system or to both systems. In any case, the irrigation system is a high hazard for several reasons. Most systems are constructed of materials that are not suitable for use with potable water. Sprinklers, bubbler outlets, emitters, and other equipment are exposed to substances such as fecal material, fertilizers, pesticides, and other chemical and biological contaminants. Sprinklers generally remain submerged under water after system use or storms. Irrigation systems can have various design and operation configurations. They may be subject to various onsite conditions, such as additional water supplies, chemical injection, booster pumps, and elevation changes. All of these conditions must be considered in determining proper backflow protection.

Solutions of chemicals and/or fertilizers are used in or around irrigation systems for many purposes. Some of the chemical compounds that may be injected or aspirated into, or come in contact with, these systems include

- Fertilizers: Ammonium salts, ammonia gas, phosphates, potassium salts
- Herbicides: 2,4-D, dinitrophenol, 2,4,5-T, T-pentachlorophenol, sodium chlorate, borax, sodium arsenate, methyl bromide
- Pesticides: TDE, BHC, lindane, TEPP, parathion, malathion, nicotine, MH, and others
- Fecal matter: Animal and other

Protection recommended. For irrigation systems connected to the public water system, the appropriate protection is an air-gap separation or a reduced-pressure principle backflow-prevention assembly.

A pressure vacuum breaker may be used for service protection if the water service is the sole source of supply to the premises or property, if it is used strictly for irrigation (such as for median islands and parking strips), and if there is no means or potential means for backpressure.

For irrigation systems connected internally to the private plumbing system, the appropriate protection is outlined in the locally adopted plumbing code. However, if the system is not protected according to the water purveyor's requirements, protection should be required at the service connection as outlined above.

Laundries and Dye Works (Commercial Laundries)

For purposes of this discussion, a laundry does not include self-service laundries, including Laundromats, except where the laundry equipment constitutes a cross-connection.

The hazards normally found in plants of this type include cross-connections between the potable water system and laundry machines with under-rim or bottom inlets; dye vats in which toxic chemicals and dyes are used; water-storage tanks equipped with pumps and recirculating systems; shrinking, bluing, and dyeing machines with direct connections to circulating systems; retention and mixing tanks (Some of these machines or equipment have pumps that can pump contaminated fluids through cross-connections into the public water supply.); sewage pumps used for priming, cleaning, flushing, or unclogging purposes; water-operated sewage sump ejectors used for operational purposes; sewer lines used for disposing of filter or softener backwash water from cooling systems, for quick draining of building lines, for flushing or blowing out obstructions, etc.; reservoirs, cooling towers, and circulating systems that may be heavily contaminated with vermin, algae, bacterial slimes, or toxic water-treatment compounds, such as pentachlorophenol, copper sulfate, chromate, metallic glucosides, compounds of mercury, and quaternary ammonium compounds; and steam-generating facilities and lines that may be contaminated with boiler compounds, such as those chemicals listed above (NOTE: A specific hazard is the possibility that steam will get back into the potable system, causing system damage or a health hazard.).

Protection recommended. An air-gap separation or a reduced-pressure principle backflow-prevention assembly is recommended.

Marine Facilities and Dockside Watering Points

The actual or potential hazards to the potable water system created by any marine facility or dockside watering point must be individually evaluated. The basic risk to a potable water system is due to the possibility that contaminated water can be pumped into the potable water system by the fire pumps or other pumps aboard ships. In addition to the normal risks peculiar to dockside watering points, risks are found at those areas where dockside watering facilities are used in connection with marine construction, maintenance and repair, and permanent or semipermanent moorages. Health authorities point out the additional risk of dockside water facilities that are located on freshwater or diluted salt water where, if backflow occurs, it can be more easily ingested because of the lack of salty taste.

Protection recommended. Minimum system protection for marine installations may be accomplished in one of the following ways:

- Where water is delivered directly to vessels for any purpose, a reduced-pressure principle backflow-prevention assembly must be installed at the pier hydrants. All hydrants in the dockside area that are used (or are available to be used) to provide water to vessels should be so protected. If an auxiliary water supply, such as a saltwater fire system, is used, the entire dockside area should be isolated from the water supplier's system by an approved air gap. Where water is delivered to marine facilities for fire protection only, and no auxiliary supply is present, all service connections should be protected by a reduced-pressure principle backflow-prevention assembly. If hydrants are available for connection to a vessel's fire system, a reduced-pressure principle backflow-prevention assembly should be installed at the user connections as well.

- Where water is delivered to a marine repair facility, a reduced-pressure principle backflow-prevention assembly should be installed at the user connection. Where water is delivered to small-boat moorages that maintain hose bibs on a dock or float, a reduced-pressure principle backflow-prevention assembly should be installed at the user connection and a hose connection vacuum breaker should be installed on each hose bib. If a sewage pump station is provided, the area should be isolated by installation of a reduced-pressure principle backflow-prevention assembly. Water used for fire protection aboard ship, connected to dockside fire hydrants, shall not be taken aboard from fire hydrants unless the hydrants are on a fire system that is separated from the domestic system by an approved reduced-pressure principle backflow-prevention assembly or unless the hydrants are protected by portable, approved reduced-pressure principle backflow-prevention assemblies.

Metal Manufacturing, Cleaning, Processing, and Fabricating Facilities

This category includes any premises where metal goods are manufactured, cleaned, processed, or fabricated, and the process involves used water and/or industrial fluids. This type of facility may be operated or maintained either as a separate function or in conjunction with a manufacturing or other facility, such as an aircraft or automotive manufacturing plant.

The hazards normally found in plants of this type include cross-connections between the potable water system and reservoirs, cooling towers, and circulating systems that may be heavily contaminated with vermin, algae, bacterial slimes, or toxic water-treatment compounds, such as pentachlorophenol, copper sulfate, chromate, metallic glucosides, compounds of mercury, and quaternary ammonium compounds; steam-generating facilities and lines that may be contaminated with boiler compounds, such as those chemicals listed above (NOTE: A specific hazard is the possibility that steam will get back into the domestic system, causing system damage or a health hazard.); industrial-fluid systems and lines containing cutting and hydraulic fluids, coolants, hydrocarbon products, glycerin, paraffin, caustic and acid solutions, etc.; plating facilities involving the use of highly toxic cyanides and heavy metals in solution, such as copper, cadmium, chrome, and nickel; acids and caustic solutions; plating-solution filtering equipment with pumps and circulating lines; tanks, vats, or other vessels used in painting, descaling, anodizing, cleaning, stripping, oxidizing, etching, passivating, pickling, dipping, or rinsing operations, or other lines or facilities needed in the preparation or finishing of the products; water-cooled equipment that may be sewer-connected, such as compressors, heat exchangers, and air-conditioning equipment; tanks, can- and bottle-washing machines, and lines where caustics, acids, detergents, and other compounds are used for cleaning, sterilizing, and flushing; hydraulically operated equipment for which the water purveyor's water pressure is directly used and may be subject to backpressure; equipment under hydraulic tests, such as tanks, lines, valves, and fittings; and pumps, pressure cylinders, or other hydraulic facilities that may be used to provide pressures for testing purposes. (NOTE: In such cases, air, gas, or hydraulic fluids may be forced back into the public system.)

Protection recommended. An air-gap separation or a reduced-pressure principle backflow-prevention assembly is recommended where there is a high hazard. A double check valve assembly should be used where there is a low hazard.

Multistoried Buildings

In terms of their internal potable water systems, multistoried buildings may be broadly grouped into the following three types: those that use only the service pressure to distribute potable water throughout the structure, with no internal potable water reservoir; those that use a booster pump to provide potable water directly to the upper floors; and those that use a booster pump to fill a covered roof reservoir, from which there is a down-feed system for the upper floors.

Considerable care must be exercised to prevent the misuse of the suction-side line to these pumps as the take-off for domestic, sanitary, laboratory, or industrial uses on the lower floors. Pollutants or contaminants from equipment supplied by take-offs from the suction-side line may easily be pumped throughout the upper floors.

In a multistoried building, regardless of the configuration of its internal potable water system, there probably are one or more take-offs for industrial water. Any loss of distribution main pressure will cause backflow from these buildings' systems unless approved backflow-prevention assemblies are properly installed.

Protection recommended. An air-gap separation or a reduced-pressure principle backflow-prevention assembly is recommended where there is a high hazard. A reduced-pressure principle backflow-prevention assembly is recommended where take-offs for sanitary facilities on lower floors are connected to the suction side of booster pump(s). A double check valve assembly should be installed on all other buildings.

NOTE: Connections to the suction side of booster pumps should be limited to avoid drawing water from adjacent unprotected premises.

Oil and Gas Production, Storage, or Transmission Properties

This category includes any premises where animal, vegetable, or mineral oils and gases are produced, developed, processed, blended, stored, refined, or transmitted in a pipeline, or where oil or gas tanks are maintained. Also included are sites where an oil well is being drilled, developed, operated, or maintained, or where an oxygen, acetylene, petroleum, or other natural or manufactured gas production or bottling plant is operated or maintained. Such premises should also include locations where oil or gas tanks, bottles, or other storage or pressure vessels are repaired, tested, or maintained; premises having dehydration or refinery facilities; premises where the water service is used for "slugging" oil or gases through transmission lines; or where the water service is used for testing or purging oil and gas tanks or oil and gas pipelines; and other similar uses.

The hazards normally found in plants of this type include cross-connections between the potable water system and steam boilers and lines; mud pumps and mud tanks; hydraulically operated "Tretolite tanks"; oil-well casings (for dampening pressures); dehydration tanks and outlet lines from storage and dehydration tanks (for purging purposes); oil and gas tanks (to create hydraulic pressures and to hydraulically raise the oil and gas levels); gas and oil lines (for testing, evacuating, and slugging purposes); reservoirs, cooling towers, and circulating systems that may be heavily contaminated with vermin, algae, bacterial slimes, or toxic water-treatment compounds, such as chromate, pentachlorophenol, copper sulfate, metallic glucosides, compounds of mercury, and quaternary ammonium compounds; steam-generating facilities and lines that may be contaminated with boiler compounds, such as those chemicals listed above (Note: A very particular hazard is the possibility that steam will get back into the domestic system, causing system damage or a health hazard.); industrial-fluid systems and lines containing cutting and hydraulic fluids, coolants, hydrocarbon products, glycerin, paraffin, caustic and acid solutions, etc.; fire-fighting systems, including storage reservoirs that may be treated for prevention of scale formation, algae, slime growths, etc.; fire systems that may be subject to contamination with antifreeze solution, liquid foam concentrates, or other chemicals or chemical compounds used in fighting fire; fire systems that are subject to contamination with auxiliary or used water supplies or industrial fluids; water-cooled equipment that may be sewer-connected, such as compressors, heat exchangers, and air-conditioning equipment; hydraulically operated equipment for which the water purveyor's water pressure is directly used and may be subject to backpressure; equipment under hydraulic tests, such as tanks, lines, valves, and fittings; and pumps, pressure cylinders, or other hydraulic facilities that may be used to provide pressures for testing purposes. (NOTE: In such cases, air, gas, or hydraulic fluids may be forced back into the public system.)

Protection recommended. An air-gap separation or a reduced-pressure principle backflow-prevention assembly is recommended.

Other Fire-Suppression Systems

Whenever the potable water supply, service line, or branch line within a building, property, or site is used to supply a fire-suppression system, there must be appropriate backflow protection. Fire-suppression systems other than wet-pipe fire sprinkler systems that are directly connected to public water main lines vary in their backflow requirements. Wet-pipe systems connected to private plumbing systems typically are regulated by local plumbing codes.

Dry-pipe pressurized and dry-pipe nonpressurized are two common examples of general categories of fire-suppression systems that are not wet-pipe systems. Some dry-pipe deluge fire-suppression systems provide, within the system or inherent in the design of the system, backflow protection, such as an air gap or vacuum breaker.

Protection recommended. Because of the wide variety of system designs, backflow protection must be based on the type of cross-connection and the degree of hazard.

If backflow protection is not in place on the water service line to the site and the water purveyor is comfortable in accepting the internal backflow protection that exists, no additional backflow protection is required. However, if no internal backflow protection exists, or if the internal protection is not adequate and service protection is not in place, the water purveyor must require appropriate backflow protection internally or on the water service line to the facility, or both.

Paper and Paper-Product Plants

This category includes any premises where a paper or paper-products plant (wet process) is operated or maintained. Paper or paper-product plants as used here means those plants where used waters, industrial fluids, and chemicals are used in the manufacturing process.

The hazards normally found in plants of this type include cross-connections between the potable water system and pulp, bleaching, dyeing, and processing facilities, which may be contaminated with toxic chemicals; reservoirs, cooling towers, and circulating systems that may be heavily contaminated with vermin, algae, bacterial slimes, or toxic water-treatment compounds, such as chromate, pentachlorophenol, copper sulfate, metallic glucosides, compounds of mercury, and quaternary ammonium compounds; steam-generating facilities and lines that may be contaminated with boiler compounds, such as those chemicals listed above (NOTE: A very particular hazard is the possibility that steam will get back into the domestic system, causing system damage or a health hazard.); industrial-fluid systems and lines containing cutting and hydraulic fluids, coolants, hydrocarbon products, glycerin, paraffin, caustic and acid solutions, etc.; water-cooled equipment that may be sewer-connected, such as compressors, heat exchangers, and air-conditioning equipment; fire-fighting systems, including storage reservoirs that may be treated for prevention of scale formation, corrosion, algae, slime growths, etc.; fire systems that may be subject to contamination with antifreeze solutions, liquid foam concentrates, or other chemicals or chemical compounds used in fighting fire; or fire systems that are subject to contamination with auxiliary or used water supplies or industrial fluids.

Protection recommended. An air-gap separation or a reduced-pressure principle backflow-prevention assembly is recommended.

Plating Plants and Facilities

This category includes any premises where there is a mechanical, chemical, or electrochemical plating or processing plant. The plating plant or facility may be operated or maintained either as a separate function or in conjunction with a manufacturing or other facility, such as an aircraft or automotive manufacturing plant. Plating as used here includes such operations as chromium, cadmium, or other plating; galvanizing; anodizing; cleaning; stripping; oxidizing; etching; passivating; and pickling.

The hazards normally found in plants of this type include cross-connections between the potable water system and plating-solution filtering equipment with pumps and circulating lines; tanks, vats, or other vessels used in painting, descaling, anodizing, cleaning, stripping, oxidizing, etching, passivating, pickling, dipping, or rinsing operations, or other lines or facilities needed in the preparation or finishing of the products; steam-generating facilities and lines that may be contaminated with boiler compounds, such as pentachlorophenol, copper sulfate, chromates, metallic glucosides, and compounds of mercury (NOTE: A specific hazard is the possibility that steam will get back into the domestic system, causing system damage or a health hazard.); and water-cooled equipment that may be sewer-connected, such as compressors, heat exchangers, and air-conditioning equipment.

Protection recommended. An air-gap separation or a reduced-pressure principle backflow-prevention assembly is recommended.

Radioactive Material or Substances, Plants or Facilities Handling

This category applies to any premises where radioactive materials or substances are processed in a laboratory or plant, where these materials may be handled in such a manner as to create a potential hazard to the water system, or where there is a reactor plant.

Protection recommended. An air-gap separation or a reduced-pressure principle backflow-prevention assembly is recommended.

Reclaimed or Recycled Water

Generally, reclaimed water is treated sewage effluent that undergoes further treatment to improve its quality. It is critical to remember that reclaimed water is nonpotable. It can contain aluminum, boron, calcium, chromium, copper, iron, lead, magnesium, manganese, molybdenum, nickel, nitrogen, phosphorus, potassium, sodium, and zinc, as well as *Cryptosporidium*, entamoeba, fecal coliforms, *Giardia*, and other parasites and bacteria. Reclaimed water use is not new; some use dates back to the 1920s. Today, reclaimed water has become a popular means to conserve potable water. Every day, more extensive reclaimed-water distribution systems are being constructed and expanded. Reclaimed water is regularly used for irrigation applications, including golf courses, parks, ball fields, schools, median islands, cemeteries, and commercial and residential landscapes. In some areas it is used for other applications, such as cooling towers, toilet flushing, dust control, and general construction purposes.

Though reclaimed-water systems are a tremendous conservation resource, they do pose new risks if they are improperly maintained, operated, or identified. To ensure public health and safety, safeguards must be put in place and maintained at all times. This would also include any property having private wells with systems supplied by them. Safeguards include the installation of proper backflow prevention and the assurance that there is not a direct cross-connection between a site's potable water system and the reclaimed-water system. Several test methods are used to ensure against existing cross-connections. All involve pressure testing in one form or another; and, in some instances, they involve using a dye that is safe for potable water. When selecting or developing a test method to ensure system separation, it is recommended that potable water be used initially to test the separation of the proposed reclaimed system. This precaution is extremely important because it prevents accidental contamination of the potable water system if an unknown cross-connection exists between the proposed new or a converted existing system to be supplied by reclaimed water and the site's potable water system.

Other precautions should be taken. Water purveyors that supply reclaimed water to customers should conduct thorough site evaluations before startup and periodically thereafter. There should be no locations on the site where ponding could occur or where reclaimed-water overspray could contact food or drinking areas. Signs specifying the use of reclaimed water should be posted at all points of entry to the site. The use of standard hose bibs on the reclaimed water system should not be permitted. All reclaimed water piping, valves, meters, controls, and equipment should be clearly labeled and marked. Reclaimed-water lines should be separated from potable water lines following the same requirements for the separation of potable water lines from sewage lines. Customer education is a must.

Protection recommended. An air-gap separation or a reduced-pressure principle backflow-prevention assembly is recommended on each potable water line entering a reclaimed-water use site.

NOTE: Where reclaimed water is used for an industrial purpose, such as cooling towers, there may be a backflow preventer on the reclaimed-water line to such equipment to prevent chemicals that are added to the equipment for maintaining pH levels, corrosion control, etc., from backflowing into the supplying reclaimed-water system. Field-test equipment used in testing these backflow-prevention assemblies should be clearly identified as nonpotable field-test equipment and should be kept separate from field-test equipment used to test backflow-prevention assemblies supplied by potable water. Additionally, the same consideration for separating the field-test equipment must be given to the equipment used to conduct regular calibration checks of the equipment.

Residential Water Services

Residential water services are a single water service providing water to premises that are solely used and constructed to provide a living place for a single individual, couple, or single family. These water services can present backflow hazards to water purveyors if proper backflow protection is not installed internally or at each water service. It is recommended that all residential water services be evaluated to establish a degree of hazard and to determine the service or internal protection required. If a water purveyor elects not to require backflow protection for residential water services, it must realize that it is accepting some degree of risk.

Some items to consider in establishing a degree of hazard and in requiring backflow protection on residential water services are pets, livestock, fish, chemicals, pools, fountains, tanks, irrigation, dialysis equipment, developing equipment, gray water, reclaimed water, an auxiliary water supply, heating and cooling equipment, and other equipment or operations that use water. The elevation of the site's plumbing system above the water service connection also should be considered. If the site does not have one of the aforementioned hazards and the plumbing system meets current plumbing code requirements, the water purveyor may elect to forego service protection.

If the water purveyor determines that conditions warrant service protection, the following backflow protection is recommended.

Protection recommended. An air-gap separation or reduced-pressure principle backflow-prevention assembly should be required on each water service line to the site when a high hazard exists. A double check valve assembly should be required on each water service line to the site when a low hazard exists.

NOTES:
- Typically, residential plumbing systems are smaller than most commercial plumbing systems. Therefore, the adverse effects of thermal expansion can be greater and more immediate. As required with any closed system, thermal-expansion protection that meets the plumbing code must be installed to ensure the safety and longevity of the private plumbing system.
- The water purveyor should actively work to educate residential customers, as well as the entire community, regarding the hazards of backflow.

Residential, Single-Family Fire Sprinkler Systems

Residential fire sprinkler systems are installed in single-family dwellings that have a main water supply line of 1.5 in. or less.

Protection recommended. A reduced-pressure principle backflow-prevention assembly or air gap is recommended where there is a high hazard. A double check valve assembly is recommended where there is a low hazard.

NOTE: It is recommended that residential fire sprinkler systems that are constructed of materials approved for potable water and are flow-through (not closed) systems do not require the installation of a backflow assembly. The ends of these systems are connected to a fixture that is regularly used. This prevents the water in the system from becoming stagnant.

Restricted, Classified, or Other Closed Facilities

This category includes a service connection to any facility that is not readily accessible for inspection by the water purveyor because of military secrecy requirements or other prohibitions or restrictions. In selecting the protection recommended, consider the potential for cross-connection to sewer systems.

Protection recommended. An air-gap separation or a reduced-pressure principle backflow-prevention assembly is recommended.

Solar Domestic Hot-Water Systems

The hazards normally found in solar domestic hot-water systems include cross-connections between the potable water system and heat exchangers, tanks, and circulating pumps. Depending on the system's design, the heat-transfer medium may vary from domestic water to antifreeze solutions, corrosion inhibitors, or gases. The degree of hazard will range from a low hazard when potable water is used to a high hazard when a toxic transfer medium, such as ethylene glycol, is used. Contamination occurs when the piping or tank walls of the heat exchanger between the potable hot water and the transfer medium begin to leak.

Liquid-to-liquid solar heat exchangers can be classified as follows:

- *Single wall with no leak protection (SW)*: A heat exchanger that provides single-wall separation between the domestic hot water and the transfer medium. Failure of this wall will result in a cross-connection between the domestic hot water and the heat-transfer medium.
- *Double wall with no leak protection (DW)*: A heat exchanger that has two separate, distinct walls separating the potable water and the transfer medium. A cross-connection between the potable hot-water system and the transfer medium requires independent failure of both walls.
- *Double wall with leak protection (DWP)*: A heat exchanger that has two separate, distinct walls separating the potable hot water and the transfer medium. If a leak occurs in one or both walls of the DWP, the transfer medium will flow to the outside of the heat exchanger, thus indicating the leak.

Protection recommended. The recommendations in the following table are to be used as a guide to recommend protection for solar domestic hot-water systems.

Hazard Rating of Transfer Medium	Heat Exchanger	Protection Recommended
Non-health	SW	DC
Non-health	DW, DWP	None*
Health	SW, DW	RP
Health	DWP	None*

*Some jurisdictions may require backflow protection and/or require all heat exchangers to be double wall with leak detection. Check local plumbing codes.

NOTE: DC = double check valve assembly; DW = double wall with no leak detection; DWP = double wall with leak detection; RP = reduced-pressure principle backflow-prevention assembly; SW = single wall with no leak detection

Steam Boiler Plants

Most boiler plants use some form of boiler feedwater treatment. The chemicals normally used for this purpose include highly toxic compounds, such as cyclohexylamine, hydrazine, morpholine, and benzylamine; and less toxic compounds, such as acids, sodium hydroxide, sodium sulfite, sodium phosphate, sodium nitrate, sodium aluminate, and sodium alginate.

Protection recommended. An air-gap separation or a reduced-pressure principle backflow-prevention assembly is recommended.

Water-Hauling Equipment

This category includes any portable or nonportable spraying or cleaning units that can be connected to any potable water supply that does not contain a built-in air gap.

The hazards normally found with water-hauling equipment include cross-connections between the potable water system and tanks contaminated with toxic chemical compounds used in spraying fertilizers, herbicides, and pesticides; water-hauling tanker trucks used in dust control; and other tanks on cleaning equipment.

Protection recommended. An air-gap separation or a reduced-pressure principle backflow-prevention assembly installed at the point of the connection supplying such equipment is recommended. Hoses or piping to the equipment may not be of potable quality or may have been in contact with contaminants. In addition, in some cases the inspection and maintenance of reduced-pressure principle backflow-prevention assemblies on portable units are questionable. A water purveyor may wish to designate specific watering points, such as those equipped with air gaps, for filling portable units. This better affords monitoring by the water purveyor.

Wet-Pipe Fire Sprinkler Systems

Wet-pipe fire sprinkler systems contain water that is connected to the potable water system. It has been shown that water contained in closed or nonflow-through fire systems may be stagnant or contaminated beyond acceptable drinking water standards. Some of the contaminants found in fire sprinkler systems are antifreeze, chemicals used for corrosion control or as wetting agents, oil, lead, cadmium, and iron. Requirements stipulated by the latest model building/plumbing codes and Occupational Safety and Health Administration (OSHA) regulations require the installation of an approved backflow assembly for all new wet-pipe sprinkler systems.

For existing wet-pipe fire sprinkler systems determined to pose only a low-hazard threat, the water purveyor may consider an alternative to installing an approved backflow-prevention assembly. This option is outlined below. This is due to the retrofit cost burden, design hydraulics, and the operational principles of a modern Underwriters' Laboratories (UL)-listed alarm check valve, which has a rubber disc.

Protection recommended. For new systems: A reduced-pressure principle backflow-prevention assembly, reduced-pressure principle detector backflow-prevention assembly, or air gap is recommended where there is a high hazard. A double check valve assembly, double check detector backflow-prevention assembly, or air gap is recommended for all other closed or nonflow-through systems.

For existing systems: An air gap, reduced-pressure principle backflow-prevention assembly, or reduced-pressure principle detector backflow-prevention assembly is recommended where there is a high hazard.

For systems presenting a low hazard only and having a modern UL-listed alarm check valve that contains no lead, it is recommended that the check valve be maintained in accordance with NFPA 25 (National Fire Protection Association Manual 25, *Standard for the Inspection, Testing, and Maintenance of Water-Based Fire*

Protection Systems). When an existing sprinkler system with an alarm check valve is significantly expanded or modified, requiring a comprehensive hydraulic analysis, a double check valve assembly should be installed. See NFPA Manual 13 for information about alarm provisions.

For existing systems that present a low hazard and that have an alarm check valve containing lead, it is recommended that a UL-classified double check valve assembly be installed. See NFPA Manual 13 for information about alarm provisions.

NOTES:

- Before installing or testing a backflow-prevention assembly on a fire sprinkler system, it is recommended that the fire authority that has jurisdiction be consulted for additional criteria it may require. Additionally, it is recommended that a thorough hydraulic analysis be performed before installing a backflow-prevention assembly on a fire system.
- It is recommended that water purveyors consider allowing an existing fire sprinkler system already equipped with a double check valve assembly on the service connection to install a reduced-pressure principle backflow-prevention assembly on an added antifreeze loop(s).
- It is recommended that water purveyors evaluate the presence, use, and potential hazard of fire department connections (FDCs) to determine whether a fire system should be classified as a high hazard because of chemicals or auxiliary water sources used by the fire authority.

RECOMMENDED PROTECTION FOR WATER PURVEYOR'S HAZARDS

Distribution System

The water purveyor's distribution system poses two general categories of cross-connection hazards. The first comprises permanent connections used for draining tanks, reservoirs, and mains; to facilitate air release and vacuum relief in mains; for fire hydrants and other appurtenances with underground drain ports; for irrigation-system connections at reservoir, well, and booster sites; and for potable backup supply to reclaimed-water reservoirs. The second category of cross-connection hazard includes temporary connections used for the direct supply of water for construction and maintenance work (e.g., disinfection of new mains); the filling of tanker trucks or trailers (e.g., construction water for dust control and water for pesticide applications or fire fighting); and flushing of sewers and storm drains.

Protection recommended. For permanent connections:

- Reservoirs and storage tanks: A screened air-gap separation is recommended on overflow pipes.
- Air-release and vacuum valves: A screened air-gap separation is recommended on air-discharge outlet pipes.
- Fire hydrants and other appurtenances with underground drain ports: Eliminate all underground drain connections wherever possible. For dry barrel fire hydrants, no recommended protection is presently available.
- Irrigation-system connections: Onsite irrigation systems should have backflow protection installed according to plumbing code.
- Backup supplies to reclaimed-water reservoirs and tanks: A screened air gap installed on the inlet pipe to the vessel is recommended.

For temporary connections:

- Supply of water for filling or disinfecting new mains, etc.: A reduced-pressure principle backflow-prevention assembly is recommended.
- Supply of water for construction sites, filling tanks, etc.: An air-gap separation or a reduced-pressure backflow-prevention assembly is recommended.
- Supply of water for sewer flushing: Because of the very high hazard and the difficulty in monitoring the maintenance of an air gap (or, where allowed, a reduced-pressure backflow-prevention assembly), it is recommended to prohibit the use of water from hydrants for sewer flushing and instead to require all sewer flushing water to be provided from tanker trucks.

Treatment Plants

Water treatment plants fall into four basic categories of health risk with respect to cross-connection control.

Group A:

- The treatment is for surface water or groundwater under the influence where the primary hazard is microbiological. The backflow of raw water into the potable water system could be an acute health hazard.
- The treatment is for a primary chemical contaminant that would be an acute health hazard if the raw water backflowed into the potable water.

Group B:

- The treatment is for a primary chemical contaminant that would be a chronic health hazard if the raw water backflowed into the potable water.
- The treatment is for a secondary chemical contaminant (e.g., removal of manganese from groundwater) that would be an aesthetic problem if the raw water backflowed into the potable water.

Group C:

- The treatment processes use chemicals that would constitute an acute health hazard if they were to backflow into the potable water. This could be different from overdosing the treated water. The potable water in the plant or in the distribution system that supplies the plant could receive the chemical in a higher concentration than the intended dose.

Group D:

- The treatment uses no chemicals (e.g., it uses an aeration and filtration system for iron removal) or it uses only chemicals (e.g., sodium hypochlorite) that would not be considered an acute health hazard if ingested.

In addition, all treatment plants should consider the cross-connection risks from the following:

- The waste product must be discharged somewhere. It could be to a holding pond or a reclaimed-water tank for reprocessing, to a backwash discharge swale or pond, or to a sewer.
- Laboratory facilities are operated for quality control. Even simple aeration and pressure filtration systems for manganese removal would use a field-test kit containing cyanide.
- Chemicals are periodically used for cleaning pipes and tanks. For example, an electrodialysis reversal system could use hydrochloric acid for clean-in-place application to the membranes.

Water treatment plant cross-connection control should provide in-premises protection for the water purveyor's employees, quality assurance for the product water, and protection for the distribution system.

Protection recommended. For in-premises protection for the water purveyor's employees, the water purveyor should

- At a minimum, comply with the plumbing and safety codes that normally govern private property, even if the water purveyor's plant is exempt from normal plumbing inspection requirements. (NOTE: The plumbing code applies to potable water systems.)
- Label all pipes in the treatment system, from the raw water source to the product-discharge pipe leaving the plant or treatment train. Label all such piping as "nonpotable water."
- Provide a limited number of dedicated and labeled outlets for potable water in the treatment plant. Provide the normal plumbing fixtures for building occupancy (e.g., toilets and sinks) in a separate building or in a building addition to the treatment plant. Provide area isolation (containment) for the potable water line into the plant and laboratory facilities.

For quality assurance, the water purveyor should

- Use backflow-prevention assemblies to protect the integrity of each process in the treatment train. This should be done even though the water piping may be labeled nonpotable and thus may not be subject to plumbing code fixture-protection requirements (see Table 5-1).

Table 5-1 Recommended protection at fixtures and equipment found in water treatment plants*

Description of Fixture, Equipment or Use	Recommended Minimum Protection
Raw water storage reservoir	Screened air gap on overflow Air gap on drain No bypass for group A category of treatment
Bulk chemical storage	Air gap on dilution water supply Provide day tank with screened air gap on overflow and air gap on drain
Filter bed/filter tank discharge	Air gap on waste discharge
Surface washer	Pressure vacuum-breaker assembly/double check valve assembly
Chemical-feed pumps	Assure discharge at point of positive pressure, and antisiphon valve Check valve at discharge point Foot valve in tank No pump primer line
Chemical-feed injectors	Check valve at discharge point Check valve at injector inlet
Saturators and dry-chemical solution tanks	Air gap on fill line Screened air gap on overflow Air gap on drain
Membrane clean-in-place systems	Provide physical disconnect
Sample lines to monitoring equipment	Air gap or atmospheric vacuum breaker Label "nonpotable water"
Hose-bib connections	Hose-bib vacuum breaker

*Recommendations for backflow prevention at fixtures or equipment within the treatment plant assumes there is a service containment or area isolation as shown in Figure 5-2 and that all water use within the plant area that contains the treatment equipment shall be considered industrial and, therefore, not for consumption until the water leaves the treatment plant.

TYPICAL HAZARDS   83

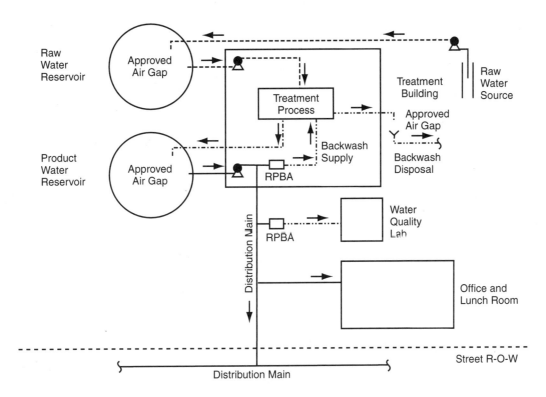

Figure 5-1 Cross-connection control, water treatment plants

For in-premises protection and quality assurance, the water purveyor should
- Use approved air gaps or testable backflow-prevention assemblies wherever possible. Periodically inspecting air gaps and testing assemblies provides significantly greater reliability.
- Provide service containment for the applicable category classification as follows (most stringent requirement applied).

Service containment based on product being treated:
  Group A: A reduced-pressure principle backflow-prevention assembly is recommended.
  Group B: A double check valve assembly is recommended.

Service containment based on chemicals used in treatment:
  Group C: A reduced-pressure principle backflow-prevention assembly is recommended.
  Group D: A double check valve assembly or fixture protection is recommended.

Service containment may not be practical in normal terms. The in-plant potable water may be supplied from the treatment system discharge. This could be the case for small, pressurized filter systems that discharge into the distribution main for transmission to a remote off-site reservoir.

Service containment is applicable where the treatment system discharges into a product storage tank for pumping through a transmission line to a remote reservoir. Pressurized water may be brought into the plant from the distribution main.

Figure 5-1 shows the general arrangement of service-containment and area-isolation backflow prevention. The water purveyor should apply to the water treatment

84 BACKFLOW PREVENTION AND CROSS-CONNECTION CONTROL

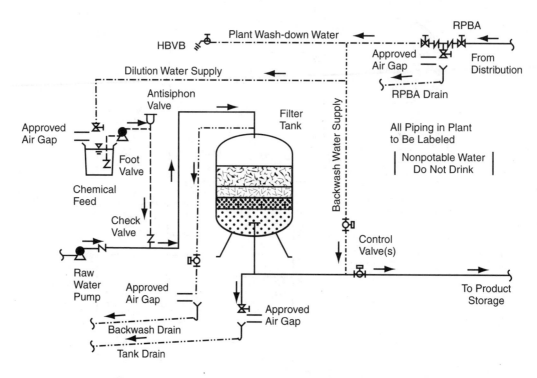

Figure 5-2 Service-containment and area-isolation water treatment plants

plant the same service-containment requirements imposed on industrial customers. Table 5-1 lists the backflow-prevention measures used to prevent cross-connections that could occur **between** individual treatment processes within the treatment train from contaminating the finished product.

Offices and Works Areas

Offices and works areas owned by the water purveyor contain the same cross-connection hazards found on customers' premises. It is important that the water purveyor follow the same cross-connection control requirements it imposes upon customers.

Protection recommended. Service protection by a containment backflow preventer following the policy established by the water purveyor for customers is recommended. In addition, compliance with the fixture-protection backflow requirements established by the plumbing code having jurisdiction is recommended.

AWWA MANUAL M14

Appendix A

Assembly Test Procedures

INTRODUCTION

Several methods may be used to test backflow-prevention assemblies. To ensure that test results obtained from certified testers are reliable, standardized testing procedures should be used. The acceptable procedures should be specified by the state or provincial health departments and the water purveyor.

The following are some of the publications or sources that provide test procedures for the reduced-pressure backflow assembly, double check valve assembly, and pressure vacuum-breaker assemblies (PVBA and/or SVBA):

- Manufacturers' literature for differential pressure gauges
- Manufacturers' literature for backflow-prevention assemblies
- *Cross Connection Control Manual*, 6th ed., Appendix F, Pacific Northwest Section of AWWA
- *Cross-Connection Control Manual*, 1989, USEPA publication 570/9-89-007
- *Backflow Prevention Assemblies—Series 5000*, 2000, American Society of Sanitary Engineering
- Field Test Procedures (Position Paper), New Englend Water Works Association
- *Manual of Cross-Connection Control*, 9th ed., Foundation for Cross-Connection Control and Hydraulic Research, University of Southern California

The test procedures listed above may vary in the test equipment used, the type of test performed to determine the operating performance of the assembly, and the criteria used to determine compliance with the operating requirements. Some tests may vary only in the sequence of the steps or in the method of troubleshooting or diagnosing problems. There are many test procedures used in different geographic regions. Many test procedures have certain idiosyncrasies that can render the data generated inaccurate. Be sure to review the many different test procedures to determine the one that provides the most accurate data most frequently.

Recommended are those test procedures based on evaluating the operating requirements specified in the design standard for assembly using a differential pressure gauge. Testing of all components of a backflow-prevention assembly to their performance criteria provides an additional safety factor, and thus, increases the reliability of an "approved assembly."

In this section of appendix A, examples of step-by-step procedures are provided for the testing of backflow-prevention assemblies. A tester should maintain the proper certification or other needed local licensing in order to test backflow preventers in the field. It is assumed that the tester is familiar with the procedures for purging test cocks, isolating the assembly, connecting test equipment, bleeding air from the test equipment, etc.

The purpose of this test procedure information is to guide the user to consider various recommended minimum performance requirements for the tester to use in testing the various types of assemblies. *No endorsement by AWWA of the various example test methods is implied.*

Included are examples from the Pacific Northwest Section manual and examples from the New England Water Works Association (NEWWA). The backflow-assembly tester must verify with the local authority the acceptability of these or any other alternate test procedure.

COMMON STEPS TO TAKE WHEN TESTING ASSEMBLIES

Prior to initiating a test of any backflow-prevention assembly, follow these procedures:[*]

1. Obtain permission from the owner, or their representative, to shut down the water supply. This is necessary to ensure that since all testing is accomplished under no-flow conditions, the owner is aware that the water supply will be temporarily shut off while the testing is being performed. Some commercial and industrial operations require constant and uninterrupted water supplies for cooling, boiler feed, and seal pump water, and water service interruptions cannot be tolerated. The water supply to hospitals and continuous process industries cannot be shut off without planned and coordinated shutdowns.

    For premise isolation assemblies, although notice can be given by the purveyor for an interruption of service, whenever possible, it is preferable to cooperate with the owner to arrange a mutually agreeable time for a shutdown.

    The request to shut down the water supply is a necessary prerequisite to protect the customer as well as limit the liability of the tester.

    Concurrent with the request for permission to shut off the water, it is advisable to point out to the owner that while the water is shut off during the test period, any inadvertent use of water within the building will reduce the water pressure to zero. Backsiphonage could result in the building's plumbing system being contaminated through cross-connections. To address this situation, it is recommended that the owner caution the inhabitants of the building not to use water until the backflow-assembly test is completed and the water pressure restored. Additional options available to the owner would be the installation of two backflow assemblies in parallel that would enable a protected bypass flow around the assembly to be tested. Also if all water outlets are properly equipped with backflow assemblies and devices within the building, backsiphonage would not occur while assemblies are being tested, or for other reasons.

---

[*]Portions copied with permission from work by Howard D. Hendrickson, P.E., Water Service Consultants. Mr. Hendrickson's work is also printed in the *Cross-Connection Control Manual*, 1989, USEPA.

2. Determine the serial number and the type of assembly to be tested, i.e., RPBA, DCVA, PVBA, or SVBA.
3. Determine the flow direction (reference directional flow arrows or wording provided by the manufacturer on the assembly).
4. Number the test cocks (mentally), flush them of potential debris, and assemble appropriate test cock adapters and bushings that may be required.
5. Shut off the downstream isolating valve (shutoff valve no. 2).
6. Hook up the test equipment in the manner appropriate to the assembly being tested and the specific test being performed.

PNWS—TEST PROCEDURES FOR REDUCED-PRESSURE BACKFLOW AND REDUCED PRESSURE DETECTOR ASSEMBLIES USING DIFFERENTIAL PRESSURE GAUGE[*]

Relief Valve

Performance criteria. During normal operating conditions, whether or not there is flow through the assembly, the pressure in the zone between the check valves (zone of reduced pressure) shall be at least 2 psi less than the pressure on the inlet (supply) side of the assembly. When there is no flow from the inlet (supply) side of the assembly and the inlet pressure drops to 2 psi, the pressure within the zone of reduced pressure shall be atmospheric. If the inlet pressure drops below 2 psi, the relief valve shall continue to open (ANSI/AWWA C511, Sec. 4.2.1; Sec. 4.2.2).

Test objective, method, and reporting requirements. The first test objective is to determine the opening point and operation of the differential pressure relief valve. To do so, the pressure between the check valves (zone of reduced pressure) must be increased by slowly bypassing water from upstream of check valve no. 1 until the differential pressure begins to decrease. This is done through the differential pressure gauge test kit (Figure A-1) by bypassing higher supply pressure from test cock no. 2 into the lower pressure of the zone of reduced pressure through test cock no. 3. Closely observe the differential pressure as it slowly drops. When the first drop of water is observed, note the differential pressure. This value is the opening point of the relief valve and must be 2.0 psid or greater (psid refers to differential pressure) (Figure A-2).

Record on the test report form the differential pressure gauge reading of the point of initial opening of the relief valve.

The second test is to verify that the relief valve will continue to open with a decrease in the differential pressure below the point which the relief valve begins to drip. This is considered an important factor in the issue of verification of the continued performance of an RPBA or RPDA. Although it is preferable to also verify that the relief valve will open fully when the supply pressure drops to atmospheric, space restriction around the assembly often makes it impractical to do so in the field test on all assemblies.

To determine that the relief valve will continue to open with a decrease in the differential pressure, the flow of water must be increased between test cock no. 2 and test cock no. 3. This may be done by fully opening the low-side control valve on the differential pressure gauge. On 2.5-in. and larger assemblies, it may be necessary to install a bypass hose separate from test equipment between test cock no. 2 and test cock no. 3 to provide a significant flow to check that the relief valve will continue to open.

---

[*]Reprinted with permission from the Pacific Northwest Section of the American Water Works Association, *Cross Connection Control Manual*, 6th ed.

88 BACKFLOW PREVENTION AND CROSS-CONNECTION CONTROL

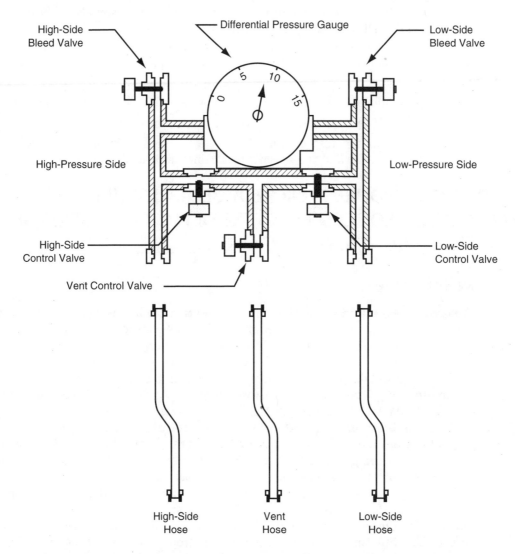

Figure A-1 Major component parts of a five-valve differential pressure gauge

If flow from the relief valve increases as an increased supply of water is bypassed into the zone of reduced pressure, the relief valve shall be considered to continue to open.

Record on the test report form that the relief valve "continued to open."

Check Valve No. 1

Performance criteria. Check valve no. 1 shall seal tight in the direction of flow at an adequate pressure to prevent the relief valve from opening and to prevent excessive discharge due to pressure fluctuation. The minimum pressure drop (differential) across check valve no. 1 under normal flow conditions shall be at least 3.0 psi greater than the pressure differential necessary to cause the relief valve to open (ANSI/AWWA C511, Sec. 4.2.1, Sec. 4.2.6).

Test objective, method, and reporting requirements. To test check valve no. 1 for tightness in the direction of flow, determine the static pressure drop across the check valve using a differential pressure gauge test kit.

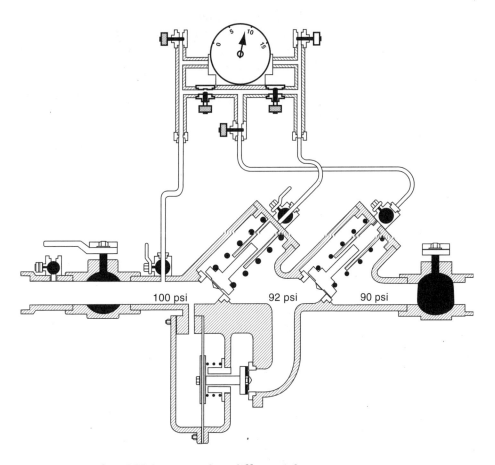

Figure A-2 Illustration of an RPBA test with a differential pressure gauge

The pressure differential gauge reading shows the "apparent" pressure drop (differential) across check valve no. 1. If the gauge reading remains steady, check valve no. 1 shall be considered to hold tight in the direction of flow. This test assumes that the relief valve operates. This test is valid only after the test of the relief valve is completed and the relief valve is confirmed to be operable. However, the test may be performed before the test of the relief valve.

Record this differential pressure gauge reading on the test report form as the check valve no. 1 pressure drop and state that check valve held tight in direction of flow.

This test does not confirm that the check valve will hold tight against backpressure. It is assumed that if the check valve holds at least 1.0 psi differential in the normal direction of flow, it will hold tight in the reverse direction of flow.

To check for the minimum 3.0 psi "buffer," subtract the pressure differential gauge reading for the relief valve to drip from the pressure drop across check valve no. 1. The actual test is under static conditions, since the pressure drop "under normal flow conditions" varies with flow rate. This value shall be 3.0 psi or greater.

Record this value on the test report form.

Check Valve No. 2

Performance criteria. Check valve no. 2 shall be internally loaded so that when the pressure on the inlet (supply) side of the valve is at least 1 psi and the outlet (downstream) pressure is atmospheric, the check valve will be drip-tight in the normal direction of flow (ANSI/AWWA C511, Sec. 4.2.5).

Test objective, method, and reporting requirements. To test check valve no. 2 for tightness in the direction of flow, determine the static pressure drop across the check valve using a differential pressure gauge test kit. The test differs from the test of check valve no. 1 in that the downstream pressure is atmospheric.

This test may be made with the differential pressure gauge high-side hose connected only to test cock no. 3 and test cock no. 4 open. The test kit must be held at the centerline of the assembly or at the elevation of test cock no. 4 if the test cock is located on the top of the check valve. After water stops flowing from test cock no. 4 and the gauge stabilizes, the differential pressure indicated by the gauge is the static pressure drop across check valve no. 2. The pressure drop must be 1.0 psid or greater.

Record this differential pressure gauge reading as the check valve no. 2 pressure drop and state that check valve held tight in direction of flow.

It is recommended that check valve no. 2 be tested for tightness in the reverse direction of flow (backpressure condition) if the above-noted direction of flow test is prevented by leaking isolating valves.

Bypass Meter on RPDA

Performance criteria. The bypass meter must register any flow (e.g., 3 to 5 gal) that occurs through the assembly (main line or bypass). However, it is not necessary that the meter accurately register the flow.

Test objective, method, and reporting requirements. Partially open the main-line assembly's test cock no. 4. Observe bypass meter; meter dial should move to register flow.

In addition, if test cock no. 4 of the main-line assembly is located on the bypass piping (rather than on the body of the main-line assembly), close shutoff valve no. 2 on the bypass assembly, partially open test cock no. 4. If flow continues from test cock, this indicates that bypass connection to the body of the main-line assembly is not restricted.

Record on test report form that detector meter registered flow.

Air Gap

Performance criteria. The distance of the air gap below the relief valve discharge vent (port) shall be in compliance with the requirements for an approved air gap.

Test objective, method, and reporting requirements. Measure the distance between the relief valve vent and the overflow rim of the receiving drain fixture.

Record on the test report form that the air gap is in compliance.

```
RPBA/RPDA
Relief Valve              Dripped at:                              __ . __ psi
[ ☐ 2.0 psid ]            or failed to open? ____ (check)
                          Continued to open?                       yes ____, no ____

Check Valve no. 1         Pressure drop:                           __ . __ psi
[ ☐ 1.0 psid ]               Valve tight?                          yes ____, no ____

Check Valve no. 1         C V no.1 pressure drop                   __ . __ psi
Buffer [ ☐ 3.0 psi ]      minus relief valve psid

Check Valve no. 2         Pressure drop:                           __ . __ psi
[ ☐ 1.0 psid ]               Valve tight,
                                Flow direction?                    yes ____, no ____
                                Backpressure?                      yes ____, no ____

Air Gap distance adequate?                                         yes ____, no ____
Test Cock no. 4 opened, meter moved?                               yes ____, no ____
Detector Meter Reading: _____
```

PNWS—TEST PROCEDURES FOR DOUBLE CHECK VALVE AND DOUBLE CHECK DETECTOR ASSEMBLIES USING DIFFERENTIAL PRESSURE GAUGE*

Check Valve No. 1 and Check Valve No. 2

Performance criteria. Check valves shall be loaded so that when the supply pressure is at least 1.0 psi and the outlet pressure is atmospheric, each check valve shall be drip-tight in the normal direction of flow. There shall be no leakage past any check valve when the pressure conditions that cause backflow are present (ANSI/AWWA C510, Sec. 4.2).

Test objective, method, and reporting requirements. To test either check valve no. 1 or check valve no. 2 for tightness in the direction of flow, determine the static pressure drop across the check valve using a differential pressure gauge test kit (Figure A-3).

Both shutoff valve no. 1 and shutoff valve no. 2 must be closed. For check valve no. 1, this test may be made with the differential pressure gauge high-side hose connected only to test cock no. 2 and test cock no. 3 open (to atmosphere). For check valve no. 2, this test may be made with the differential pressure gauge high-side hose connected only to test cock no. 3 and test cock no. 4 open. For a valid pressure gauge reading, the

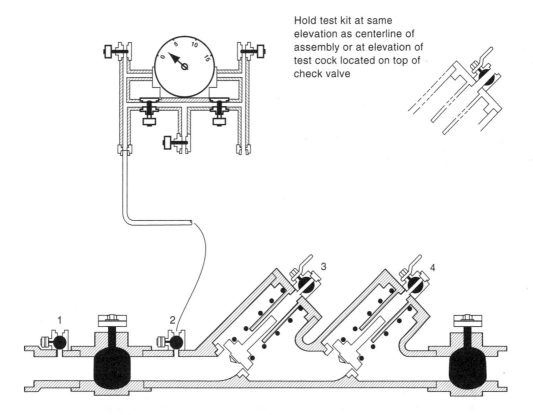

Figure A-3 Illustration of a DCVA test with a differential pressure gauge

---

*Reprinted with permission from the Pacific Northwest Section of the American Water Works Association, *Cross Connection Control Manual*, 6th ed.

test kit must be held at the centerline of the assembly or at the elevation of test cock no. 4 (or test cock no. 3, for testing check valve no. 1) if the test cock is located on the top of the check valve. After water stops flowing from test cock no. 4 and the gauge stabilizes, the differential pressure indicated by the gauge is the static pressure drop across the check valve. The pressure drop must be 1.0 psid or greater.

Record this differential pressure gauge reading on the test report form as the check valve no. 1 or check valve no. 2 pressure drop and state that check valve held tight in direction of flow.

It is recommended that check valve no. 2 be tested first to prevent entrapped air from giving an inaccurate test of check valve no. 1.

The second operating requirement is that there shall be no leakage past any check valve when the pressure conditions that cause backflow are present. It is assumed that if the check valve holds at least 1.0 psi differential in the normal direction of flow, it will hold tight in the reverse direction of flow.

Bypass Meter on DCDA

Performance criteria. The bypass meter shall register any flow that occurs through the assembly (main line or bypass). However, it is not necessary that the meter accurately register the flow.

Test objective, method, and reporting requirements. Partially open the main line assembly's test cock no. 4. Observe bypass meter; meter dial should move to register flow.

In addition, if test cock no. 4 of the main-line assembly is located on the bypass piping (rather than on the body of the main-line assembly), close shutoff valve no. 2 on the bypass assembly, partially open test cock no. 4. If flow continues from test cock, this indicates that bypass connection to the body of the main line assembly is not restricted.

Record on test report form that detector meter registered flow.

DCVA/DCDA

Check Valve no. 2	Pressure drop:	_ _ . _ psi
[☐ 1.0 psid]	Valve Tight	
	Flow direction?	yes ___, no ___
Check Valve no. 1	Pressure drop:	_ _ . _ psi
[☐ 1.0 psid]	Valve Tight	
	Flow direction?	yes ___, no ___
Test Cock no. 4 opened, metered moved?		yes ___, no ___
Detector Meter Reading: _____		

PNWS—TEST PROCEDURES FOR PRESSURE VACUUM BREAKER AND SPILL-RESISTANT VACUUM-BREAKER ASSEMBLIES USING DIFFERENTIAL PRESSURE GAUGE[*]

Air Inlet

Performance criteria. The air-inlet valve shall be open when the differential pressure in the body is no less than 1.0 psi above atmospheric pressure. The air-inlet valve shall also be fully open when the water has drained from the body.

---

[*]Reprinted with permission from the Pacific Northwest Section of the American Water Works Association, *Cross Connection Control Manual*, 6th ed.

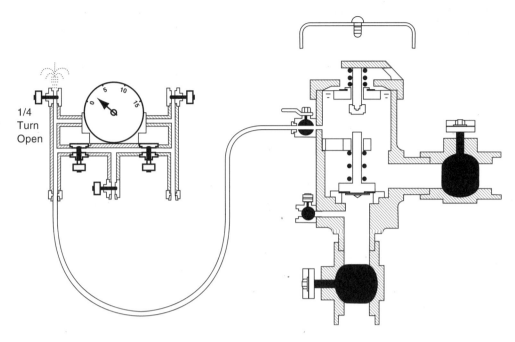

Figure A-4   Illustration of a PVBA test with a differential pressure gauge

Test objective, method, and reporting requirements. To determine the opening point of the air inlet of the PVBA, use a differential pressure gauge test kit. Both shutoff valves no. 1 and no. 2 must be closed. The differential pressure gauge high-side hose should be connected only to test cock no. 2 (Figure A-4). For a valid pressure gauge reading, the test kit must be held at the centerline of test cock no. 2. Slightly open the high-side bleed valve while observing the air inlet. Observe the differential pressure at which the air-inlet valve opens. This value must be **1.0 psid or greater**.

Record this pressure reading on the test report form.

To determine the opening point of the air inlet of the SVBA, using a differential pressure gauge test kit, follow the above method, but slightly open the air bleed screw rather than the high-side bleed valve of the differential pressure gauge.

PVBA/SVBA		
Check Valve no. 1	Pressure drop:	__ __ . __ psi
[☐ 1.0 psid]	Valve Tight?	yes ___, no ___
Air Inlet	Opened at:	__ __ . __ psi
[☐ 1.0 psid]	Air Inlet opened	yes ___, no ___

Check Valve

Performance criteria. The static pressure drop across the check valve shall be at least 1.0 psi.

Test objective, method, and reporting requirements. Slightly open the air bleed screw rather than test cock no. 2.

Equipment Description:

- Differential pressure gauge: 0–15 psid (0.1 or 0.2 psid graduations)
- Three 6-ft lengths; minimum ¼ inch I.D. high-pressure hose with screw-type couplings

- ¼-inch needle valves, for fine control of flows
- Appropriate adapter fittings for connection to various size test cocks

NEWWA—USE OF BYPASS HOSE IN REDUCED-PRESSURE PRINCIPLE BACKFLOW-PREVENTION DEVICE TESTING*

Reason. Since the second check test on a reduced pressure device consists of backpressuring the second check, a leaking downstream shutoff indicates that a flow condition exists. This flow can prevent the ability to achieve sufficient backpressure against the second check assembly, and in addition, can compromise both the first check differential reading *and* the relief valve opening point.

Procedure (knowing that the downstream shutoff is leaking). Assemble a "special fitting with built-in shutoffs" in test cock no. 4 (Figure A-5) that enables connection of both the vent hose from the test kit *and* a special, separate bypass hose connection from test cock no. 1 to test cock no. 4 (Figure A-6).

Hook up the test kit hoses in the same manner as the second check test procedure, and connect a bypass hose from test cock no. 1 to test cock no. 4. Open test cock no. 1 and introduce water into test cock no. 4.

Perform the downstream shutoff test (close test cock no. 2 on the device) (Figure A-7) and determine if sufficient water coming through the bypass hose has compensated for the water leaking through the downstream shutoff. Compensation will show up with no decay (fall-off) of the test kit needle. A satisfactory second check test can then be made because backpressuring of the second check has been achieved.

If the test cock needle still falls off, insufficient water coming through the bypass hose is available to make up for the downstream shutoff leak. Repair or replacement of the downstream shutoff may be required to achieve positive shutoff and result in accurate test results.

Position. The use of the bypass hose is only recommended when the downstream shutoff is verified as "leaking" and an additional downstream shutoff is not available, or conditions cannot be achieved, to enable testing in a verifiable no-flow condition.

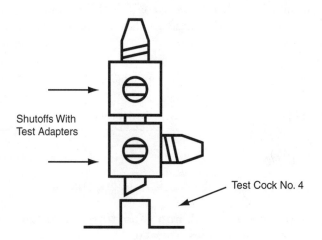

Figure A-5 Shutoff with test adaptors for use on test cock

---

*Reprinted with permission from the New England Water Works Association, Field Testing Procedure (Position Paper).

APPENDIX A   95

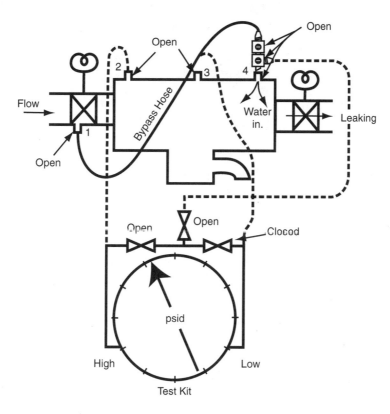

Figure A-6   Use of bypass hose

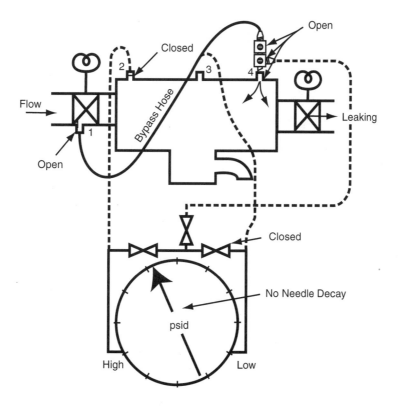

Figure A-7   Use of bypass hose for downstream shutoff test

96 BACKFLOW PREVENTION AND CROSS-CONNECTION CONTROL

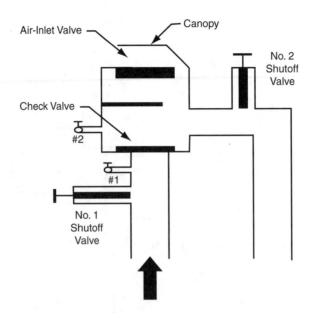

Figure A-8 Illustration of pressure vacuum breaker

NEWWA—PRESSURE VACUUM BREAKER*

The following field-testing procedure is currently used by NEWWA (Figure A-8).

Purpose of the Test

1. To determine if the static pressure across the first check is 1 psid or greater.
2. To determine if the air-inlet valve opens at least 1 psid above atmospheric pressure.

Test equipment required. A differential pressure gauge with a minimum range of 0–15 psid is required.

Field testing of a pressure vacuum breaker is accomplished utilizing a differential pressure gauge. For illustrative purposes a facsimile is shown.

Step 1 (hood removed and test adapters assembled to test cocks no. 1 and no. 2). To determine if the static pressure across the first check is 1 psid or greater (Figure A-9).

1. Close shutoff no. 2.
2. Verify that shutoff no. 1 is open.
3. Hook up the test kit high hose to test cock no. 1.
4. Hook up the test kit low hose to test cock no. 2. Place the test kit vent hose into a bucket (or suitable drainage area).
5. Position the test kit needle valves as follows:
   a. Close A.
   b. Close the needle valve at B.
   c. Open C.

---

*Reprinted with permission from the New England Water Works Association, Field Testing Procedure (Position Paper).

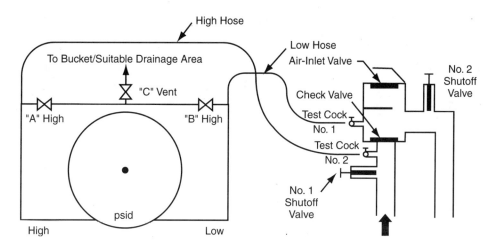

NOTE: For this test procedure, the test kit MUST be held at the same level as the assembly being tested.

Figure A-9   Illustration of pressure vacuum breaker with differential pressure gauge

    6. Open both test cock no. 1 and test cock no. 2.

    7. Open test kit needle valve A and bleed water through the vent hose into a bucket.

    8. *Slowly* close test kit needle valve A.

    9. Open test kit needle valve B and bleed water through the vent hose into a bucket.

    10. *Slowly* close test kit needle valve: It should read 1 psid or greater. A reading of less than 1 psid is cause for failure.

    11. Record data.

Shut off test cocks no. 1 and no. 2, open the test needle valves A and B, and remove the two hoses from the device. Continue to step 2.

Step 2 (with no. 2 shutoff still closed and no. 1 shutoff still open).   To determine if the air-inlet valve opens at least 1 psid above atmospheric pressure (Figure A-10).

    1. Attach the high hose to test cock no. 2. Place both the low hose and vent in the bucket.

    2. Position the test needle valves as follows:

       a. Open A.

       b. Close the needle B.

       c. Close C.

    3. Open test cock no. 2. The test kit needle should "peg" to the extreme right of the gauge face.

    4. Open test kit needle C to bleed air.

    5. Close test kit needle C.

    6. Close no. 1 shutoff.

    7. Hold up the test kit to the same level (even) with the device.

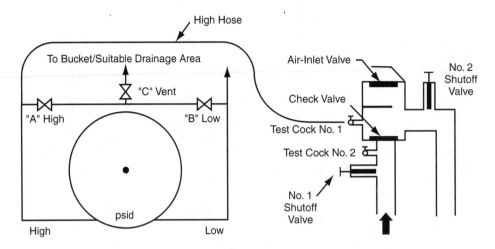

Figure A-10 Illustration of pressure vacuum breaker with differential pressure gauge (step 2)

8. *Slowly* open the test kit needle valve B while simultaneously closely observing the air-inlet valve. (Lightly placing a finger on top of the air inlet may be helpful to determine the opening point.)
9. At the point where the air-inlet valve opens (pops), read the test kit needle. It should be equal to or greater than 1 psid. A reading less than 1 psid is cause for failure.
10. Record data.

Shut off test cock no. 2, open the test kit needle valves A and B (to drain), and remove the hose. Replace the hood. Re-open shutoffs no. 1 and no. 2.

NEWWA—DOUBLE CHECK VALVE ASSEMBLY*

The following field-testing procedure is currently used by NEWWA.

It is assumed that prior to initiating a test, the following preliminary testing procedures will have been followed:

1. The type of device to be tested has been correctly determined.
2. The direction of flow has been obtained.
3. The test cocks have been numbered.
4. Test adapters have been assembled and "blown out."
5. Permission to shut down the water supply has been obtained.
6. The downstream shutoff valve has been shut off.

We will be checking the double check valve assembly for the following performance characteristics:

1. The first check valve is tight and has a minimum pressure differential across it of 1 psi.
2. The second valve is tight and has a minimum pressure differential across it of 1 psi.

---

*Reprinted with permission from the New England Water Works Association, Field Testing Procedure (Position Paper).

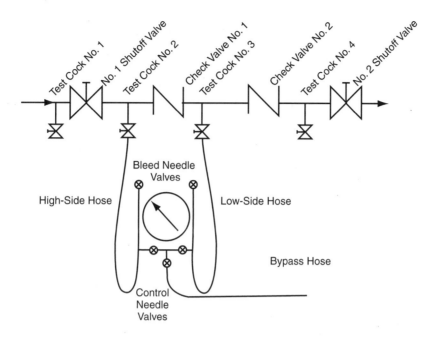

Figure A-11 Illustration of a double check valve assembly with differential pressure gauge

Test Procedure

Field testing of a double check valve assembly is accomplished utilizing a differential pressure gauge.

Step 1. Test to ensure that the first check valve is tight and has a minimum pressure differential across it of 1 psi (Figure A-11).

1. Verify that no. 1 shutoff valve is open. Close no. 2 shutoff valve.
2. Close high-pressure and low-pressure control valves on the test kit. Leave the bypass or drain hose test kit valve open.
3. Connect the high-pressure hose to test cock no. 2.
4. Connect the low-pressure hose to test cock no. 3.
5. Open test cocks no. 2 and no. 3.
6. Open the high-side bleed needle valve on the test kit, bleeding the air from the high-pressure hose. Close the high-side bleed needle valve.
7. Open the low-side bleed needle valve on the test kit, bleeding the air from the low-pressure hose. Close the low-side bleed needle valve.
8. Record the differential pressure gauge reading. It should be a minimum of 1 psid.
9. Close test cocks no. 2 and no. 3 and disconnect hoses.

Step 2. Test to ensure the second check valve is tight and has a minimum pressure differential across it of 1 psi (Figure A-12).

1. Close high-pressure and low-pressure control valves on the test kit. Leave bypass or drain hose control valve on the test kit open.
2. Connect the high-pressure hose to test cock no. 3.

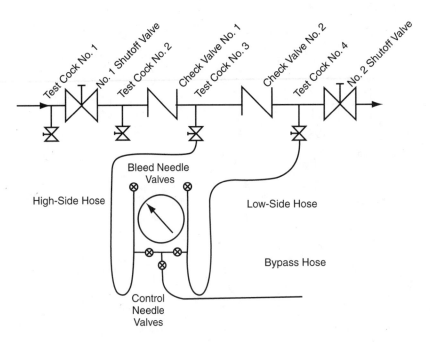

Figure A-12 Illustration of a double check valve assembly with differential pressure gauge (step 2)

3. Connect the low-pressure hose to test cock no. 4.
4. Open test cocks no. 3 and no. 4.
5. Open the high-side bleed needle valve on the test kit, bleeding the air from the high-pressure hose. Close the low-side bleed needle valve.
6. Open the low-side bleed needle valve on the test kit, bleeding the air from the low-pressure hose. Close the low-side bleed needle valve.
7. Record the differential pressure gauge reading. It should be a minimum of 1 psid.
8. Close tests cocks no. 3 and no. 4 and disconnect the hoses.

Test No. 2 Shutoff Valve for Tightness

To test the no. 2 shutoff and ensure that it is tight, and providing a positive shutoff, both check valves must be tight and holding a minimum of 1 psid. Also, little or no fluctuation of inlet supply pressure can be tolerated (Figure A-13).

1. Close high-pressure and low-pressure control valves on the test kit. Leave the bypass or drain hose control valve on the test kit valve.
2. Connect the high-pressure hose to test cock no. 2.
3. Connect the low-pressure hose to test cock no. 3.
4. Open test cocks no. 2 and no. 3.
5. Open the high-side bleed needle valve on the test kit, bleeding air from the high-pressure hose. Close the high-side bleed needle valve.
6. Open the low-side bleed needle valve on the test kit, bleeding air from the low-pressure hose. Close the low-side bleed needle valve.

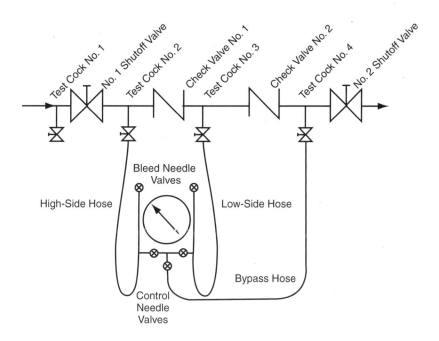

Figure A-13 Illustration of a double check valve assembly with differential pressure gauge—test for no. 2 shutoff valve test

7. Ensure that the differential pressure gauge reading is a minimum of 1 psid.
8. Connect the bypass hose to test cock no. 4 and open test cock no. 4.
9. Open the high-pressure control valve and the bypass hose control valve on the test kit. (This supplies high-pressure water downstream of check valve no. 2.)
10. Close test cock no. 2. (This stops the supply of high-pressure water downstream of check valve no. 2.)
11. If the differential pressure gauge reading holds steady, the no. 2 shutoff valve is recorded as being tight. If the differential pressure gauge reading drops to zero, the no. 2 shutoff valve is recorded as leaking.

With a leaking no. 2 shutoff valve, the device is, in most cases, in a flow condition and the previous readings taken are invalid. Unless a non-flow condition can be achieved, either through the operation of an additional shutoff valve downstream or the use of a compensating temporary bypass hose, accurate test results will not be achieved.

This completes the standard field test for a double check valve assembly. Before removal of the test equipment, the tester should ensure that all test cocks have been closed, and the no. 2 shutoff valve has been opened, thereby reestablishing flow. Also, the test kit should be thoroughly drained of all water to prevent freezing by opening all control needle valves and bleed needle valves.

All test data should be recorded on appropriate forms.

It should be noted that this test procedure tests the tightness of both the first and second check valves in the direction of flow. It does not backpressure test either check valve, and by so doing, weak or broken check springs can be reliably detected.

Extensive field testing has proven that a fouled first or second check valve, in a double check valve assembly, can seal tightly against backpressure and appears satisfactory to the tester. The backpressure test may cause the disk rubber to seat around the foul. These same fouled check valves will fail a tightness test when tested in the direction of flow. Testing for tightness and the minimum pressure differential of 1 psi in the direction of flow is a true indication of the tightness of the check valve assemblies. It also ensures proper spring tension and correct alignment of the faces of the disk rubber and seat.

It should also be noted that this test does not, under normal circumstances, require closing of the no. 1 shutoff valve. Closing the no. 1 shutoff valve may introduce debris into the device, such as rust and tuberculin common to many cast-iron gate valves. This debris could foul one or both check valves, compromising the ability of the device to protect against backflow. For this reason, NEWWA feels that closing the no. 1 shutoff valve should be avoided whenever possible. However, when encountering conditions of fluctuating supply pressure, it may become necessary to close this valve to obtain accurate test results.

NEWWA—REDUCED-PRESSURE PRINCIPLE BACKFLOW PREVENTER*

The following field-testing procedure is currently used by NEWWA.

It is assumed that prior to initiating a test, the following preliminary testing procedures will have been followed:

1. The type of device to be tested has been correctly determined.
2. The direction of flow has been obtained.
3. The test cocks have been numbered.
4. Test adapters have been assembled and "blown out."
5. Permission to shut down the water supply has been obtained.
6. The downstream shutoff valve has been shut off.
7. No water is discharging from the relief valve opening.

We will be checking the reduced-pressure principle backflow preventer for the following four performance characteristics:

1. The first check valve is tight and has a minimum pressure differential across it of 5 psi.
2. The second check valve is tight against backpressure.
3. The downstream shutoff valve is tight, providing a positive shutoff.
4. The relief valve opens at a minimum pressure of 2 psi below the inlet supply pressure.

Test Procedure

Field testing of a reduced-pressure principle backflow preventer is accomplished utilizing a differential pressure gauge.

Step 1. Test to ensure that the first check valve is tight and has a minimum pressure differential across it of 5 psi (Figure A-14).

1. Verify that no. 1 shutoff valve is open. Close no. 2 shutoff valve. If there is no drainage from the relief valve, it is assumed that the first check is tight.

---

*Reprinted with permission from the New England Water Works Association, Field Testing Procedure (Position Paper).

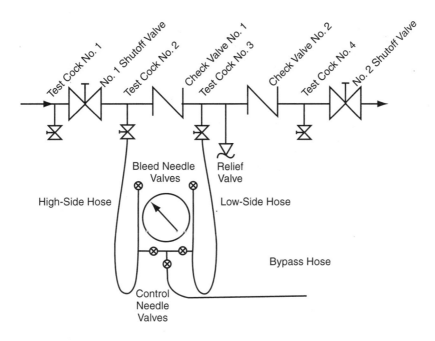

Figure A-14 Illustration of reduced-pressure principle backflow preventer test (step 1)

2. Close high-pressure and low-pressure control valves on the test kit. Leave the bypass or drain hose test kit valve open.
3. Connect the high-pressure hose to test cock no. 2.
4. Connect the low-pressure hose to test cock no. 3.
5. Open test cocks no. 2 and no. 3.
6. Open the high-side bleed needle valve on the test kit, bleeding the air from the high-pressure hose. Close the high-side bleed needle valve.
7. Open the low-side bleed needle valve on the test kit, bleeding the air from the low-pressure hose. Close the low-side bleed needle valve.
8. Record the differential pressure gauge reading. It should be a minimum of 5 psi.

Step 2. Test to ensure that the second check valve is tight against backpressure.

1. Leaving the hoses hooked up as at the conclusion of step 1, connect the bypass hose to test cock no. 4 and open test cock no. 4 (Figure A-15).
2. Open the high-pressure control valve and the bypass hose control valve on the test kit. (This supplies high-pressure water downstream of check valve no. 2.)
3. If the differential pressure gauge reading remains steady, and no water discharges from the relief valve, the second check valve is considered tight. If the differential pressure gauge reading drops and water discharges from the relief valve, the second check is recorded as leaking.

Step 3. Test to ensure that the no. 2 shutoff valve is tight, providing a positive shutoff.

1. Leaving the hoses hooked up as at the conclusion of step 2, close test cock no. 2. (This stops the supply of high-pressure water downstream of check valve no. 2.)

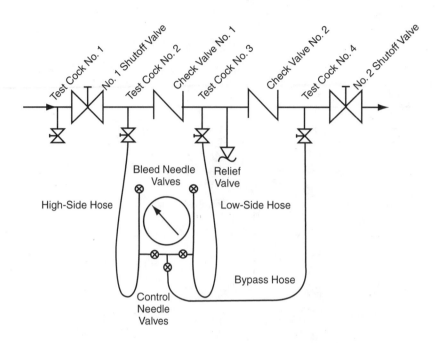

Figure A-15   Illustration of reduced-pressure principle backflow preventer test (step 2)

2. If the differential pressure gauge reading remains steady, the no. 2 shutoff valve is recorded as being tight. If the differential pressure gauge drops to zero, the no. 2 shutoff valve is recorded as leaking.

With a leaking no. 2 shutoff valve, the device is, in most cases, in a flow condition and the previous readings are invalid. Unless a nonflow condition can be achieved, either through the operation of an additional shutoff valve downstream or the use of a compensating temporary bypass hose (Figure A-16), accurate results will not be achieved.

Step 4.   To ensure that the relief valve opens at a minimum pressure of 2 psi below the inlet supply pressure.

1. With the hoses hooked up as at the conclusion of step 3, open test cock no. 2.
2. *Slowly* open the low-pressure control needle valve on the test kit.
3. Record the differential pressure gauge reading at the point when water initially drops from the relief valve opening. The differential pressure gauge reading should be a minimum of 2 psid.

This completes the standard field test for a reduced-pressure principle backflow preventer. Before removing the test equipment, the tester should ensure that the test cocks have been closed and that the no. 2 shutoff valve has been opened, thereby reestablishing flow. Also, the test kit should be thoroughly drained of all water to prevent freezing. This is done by opening all control needle valves and bleed needle valves.

All test data should be recorded on appropriate forms.

It should be noted that in this test procedure, the test of the second check valve consists of a backpressure test. If the second check valve is fouled, the no. 2 shutoff valve is tight, and the relief valve is operating properly, water will discharge from the relief valve and the differential pressure gauge reading will drop simultaneously. This constitutes ample indication that the second check valve is fouled and is a true

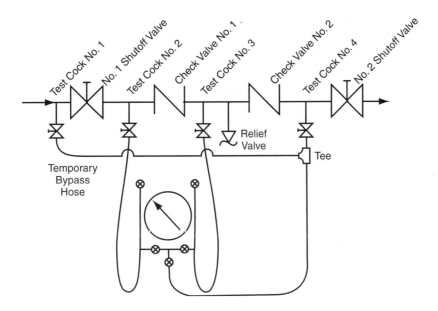

Figure A-16 Illustration of reduced-pressure principle backflow preventer test (steps 3 and 4)

representation of actual field conditions with a fouled second check valve in a reduced-pressure principle backflow preventer.

If desired, the actual pressure differential across the second check valve can be easily obtained (in the direction of flow). However, it is not felt that testing the second check valve in the direction fo flow has significant merit for standard field testing procedures over the back pressure test described here.

NEWWA—SPILL-RESISTANT PRESSURE VACUUM BREAKER*

The following field-testing procedure is currently used by NEWWA.

Purpose of the test:

1. To ensure that the differential pressure across the internal spring-loaded first check is 1 psid or greater.
2. To ensure that the air-inlet vent will start to open when supply pressure is at 1.0 psid or greater (Figure A-17).

Test equipment required:
Differential pressure gauge with a minimum range of 0–15 psid.

NOTE: For both of the following tests, the test kit must be held at the same level as the device being tested.

Test 1: To determine if the differential across the first check is at least 1 psid (Figure A-18).

1. Remove the hood.
2. Install the high hose (from A) to the test cock on the unit being tested.

---

*Reprinted with permission from the New England Water Works Association, Field Testing Procedure (Position Paper).

106 BACKFLOW PREVENTION AND CROSS-CONNECTION CONTROL

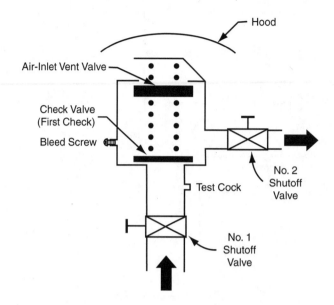

Figure A-17   Illustration of SVBA

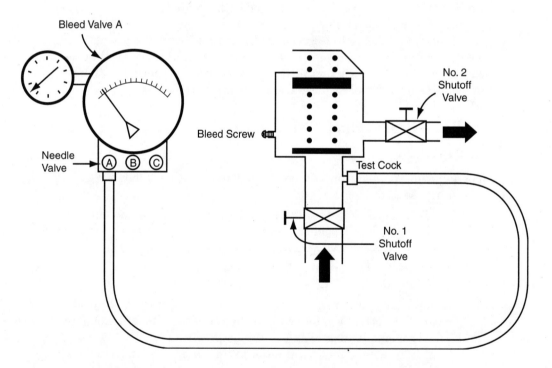

Figure A-18   Illustration of an SVBA test with a differential pressure gauge

3. Open the test cock and then open bleed valve A on top of the test kit. Bleed air from the hose, then close bleed valve A on top of the test kit.
4. Open needle valve A on the test kit.
5. Close shutoff valve no. 2.
6. Close shutoff valve no. 1.
7. *Slowly* unscrew the bleed screw.
8. When dripping from the bleed screw stops, and the needle on the test kit stabilizes, record the differential pressure. *It must be 1 psid or greater.*

Test 2: To determine at what point the air-inlet valve opens. (It must open when the inlet pressure is no less than 1.0 psi above atmospheric pressure.)

1. With both shutoffs no. 1 and no. 2 open, attach the high-side hose (from A) to the test cock on the unit being tested.
2. Open the test cock and then open the bleed valve A on top of the test kit. Bleed air from the hose, then close bleed valve A on top of the test kit.
3. Close no. 2 shutoff valve.
4. Close no. 1 shutoff valve.
5. Open the bleed screw to lower the valve outlet pressure to atmospheric.
6. Very slowly open bleed valve A on top of test kit and record the pressure reading on the gauge when the air-inlet valve opens. The reading must be 1.00 psi or greater. Air vent should open approximately $1/32$ in.

After completing above, test is over. Remove hose, reassemble hood, open no. 1 and no. 2 shutoffs, and drain all water from test kit and hose.

This page intentionally blank.

AWWA MANUAL M14

Appendix B

Model Ordinance

This model ordinance should be reviewed by the user's legal counsel prior to use.
**An Ordinance for the
Control of Backflow and Cross-Connections
Amendments to (local or state/provincial authority)
Code of (city or state/province)
Chapter _____ Section _____**

SECTION 1 CROSS-CONNECTION CONTROL—GENERAL POLICY

1.1 Purpose

The purpose of this ordinance is

1.1.1 To protect the public potable water supply of (*city, county, state, or province*) from the possibility of contamination or pollution by isolating within the customer's internal distribution system(s) or the customer's private water system(s) such contaminants or pollutants that could backflow into the public water system; and,

1.1.2 To promote the elimination or control of existing cross-connections, actual or potential, between the customer's in-plant potable water system(s) and nonpotable water systems, plumbing fixtures, and industrial piping systems; and,

1.1.3 To provide for the maintenance of a continuing program of cross-connection control that will systematically and effectively prevent the contamination or pollution of all potable water systems.

1.2 Responsibility

The (*water commissioner or state or provincial health official*) shall be responsible for the protection of the public potable water distribution system from contamination or pollution due to the backflow of contaminants or pollutants through the water service connection. If, in the judgment of said (*water commissioner or health official*) an approved backflow-prevention assembly is required (at the customer's water service connection; or, within the customer's private water system) for the safety of the water system, the (*water commissioner or health official*) or his/her designated agent shall give notice in writing to said customer to install such an approved backflow-prevention assembly(s) at specific location(s) on his/her premises. The customer shall immediately install such approved assembly(s) at his/her own expense; and, failure, refusal, or inability on the part of the customer to install, have tested, and maintain said assembly(s) shall constitute grounds for discontinuing water service to the premises until such requirements have been satisfactorily met.

SECTION 2 DEFINITIONS

2.1 Water Commissioner or Health Official

The (*commissioner or health official*) in charge of the (*water department or health department*) of the (*city, county, state, or province*) is invested with the authority and responsibility for the implementation of an effective cross-connection control program and for the enforcement of the provisions of this ordinance.

2.2 Approved

Accepted by the authority responsible as meeting an applicable specification stated or cited in this ordinance or as suitable for the proposed use.

2.3 Auxiliary Water Supply

Any water supply on or available to the premises other than the purveyor's approved public water supply. These auxiliary waters may include water from another purveyor's public potable water supply or any natural source(s), such as a well, spring, river, stream, harbor, and so forth; used waters; or industrial fluids. These waters may be contaminated or polluted, or they may be objectionable and constitute an unacceptable water source over which the water purveyor does not have sanitary control.

2.4 Backflow

The undesirable reversal of flow in a potable water distribution system as a result of a cross-connection.

2.5 Backpressure

A pressure, higher than the supply pressure, caused by a pump, elevated tank, boiler, or any other means that may cause backflow.

2.6 Backsiphonage

Backflow caused by negative or reduced pressure in the supply piping.

2.7 Backflow Preventer

An assembly or means designed to prevent backflow.

2.7.1 Air gap. The unobstructed vertical distance through the free atmosphere between the lowest opening from any pipe or faucet conveying water or waste to a tank, plumbing fixture, receptor, or other assembly and the flood level rim of the receptacle. These vertical, physical separations must be at least twice the diameter of the water supply outlet, never less than 1 in. (25 mm).

2.7.2 Reduced-pressure backflow-prevention assembly. The approved reduced-pressure principle backflow-prevention assembly consists of two independently acting approved check valves together with a hydraulically operating, mechanically independent pressure differential relief valve located between the check valves and below the first check valve. These units are located between two tightly closing resilient-seated shutoff valves as an assembly and equipped with properly located resilient-seated test cocks.

2.7.3 Double check valve assembly. The approved double check valve assembly consists of two internally loaded check valves, either spring-loaded or internally weighted, installed as a unit between two tightly closing resilient-seated shutoff valves and fittings with properly located resilient-seated test cocks. This assembly shall only be used to protect against a non-health hazard (that is, a pollutant).

2.8 Contamination

An impairment of a potable water supply by the introduction or admission of any foreign substance that degrades the quality and creates a health hazard.

2.9 Cross-Connection

A connection or potential connection between any part of a potable water system and any other environment containing other substances in a manner that, under any circumstances would allow such substances to enter the potable water system. Other substances may be gases, liquids, or solids, such as chemicals, waste products, steam, water from other sources (*potable or nonpotable*), or any matter that may change the color or add odor to the water.

2.10 Cross-Connection—Controlled

A connection between a potable water system and a nonpotable water system with an approved backflow-prevention assembly properly installed and maintained so that it will continuously afford the protection commensurate with the degree of hazard.

2.11 Cross-Connection Control by Containment

The installation of an approved backflow-prevention assembly at the water service connection to any customer's premises, where it is physically and economically unfeasible to find and permanently eliminate or control all actual or potential cross-connections within the customer's water system; or it shall mean the installation of an approved backflow-prevention assembly on the service line leading to and supplying a portion of a customer's water system where there are actual or potential cross-connections that cannot be effectively eliminated or controlled at the point of the cross-connection.

2.12 Hazard, Degree of

The term is derived from an evaluation of the potential risk to public health and the adverse effect of the hazard upon the potable water system.

2.12.1 Hazard—health. A cross-connection or potential cross-connection involving any substance that could, if introduced into the potable water supply, cause death or illness, spread disease, or have a high probability of causing such effects.

2.12.2 Hazard—plumbing. A plumbing-type cross-connection in a consumer's potable water system that has not been properly protected by an approved air gap or an approved backflow-prevention assembly.

2.12.3 Hazard—non-health. A cross-connection or potential cross-connection involving any substance that generally would not be a health hazard but would constitute a nuisance or be aesthetically objectionable, if introduced into the potable water supply.

2.12.4 Hazard—system. An actual or potential threat of severe damage to the physical properties of the public potable water system or the consumer's potable water system or of a pollution or contamination that would have a protracted effect on the quality of the potable water in the system.

2.13 Industrial-Fluids System

Any system containing a fluid or solution that may be chemically, biologically, or otherwise contaminated or polluted in a form or concentration, such as would constitute a health, system, pollution, or plumbing hazard, if introduced into an approved water supply. This may include, but not be limited to, polluted or contaminated waters; all types of process waters and used waters originating from the public potable water system that may have deteriorated in sanitary quality; chemicals in fluid form; plating acids and alkalies; circulating cooling waters connected to an open cooling tower; and/or cooling towers that are chemically or biologically treated or stabilized with toxic substances; contaminated natural waters, such as wells, springs, streams, rivers, bays, harbors, seas, irrigation canals or systems, and so forth; oils, gases, glycerin, paraffins, caustic and acid solutions, and other liquid and gaseous fluids used in industrial or other purposes for fire-fighting purposes.

2.14 Pollution

The presence of any foreign substance in water that tends to degrade its quality so as to constitute a non-health hazard or impair the usefulness of the water.

2.15 Water—Potable

Water that is safe for human consumption as described by the public health authority having jurisdiction.

2.16 Water—Nonpotable

Water that is not safe for human consumption or that is of questionable quality.

2.17 Service Connection

The terminal end of a service connection from the public potable water system, that is, where the water purveyor loses jurisdiction and sanitary control over the water at its point of delivery to the customer's water system. If a meter is installed at the end of the service connection, then the service connection shall mean the downstream end of the meter. There should be no unprotected takeoffs from the service line ahead of any meter or backflow-prevention assembly located at the point of delivery to the customer's water system. Service connection shall also include water service connection from a fire hydrant and all other temporary or emergency water service connections from the public potable water system.

2.18 Water—Used

Any water supplied by a water purveyor from a public potable water system to a consumer's water system after it has passed through the point of delivery and is no longer under the sanitary control of the water purveyor.

SECTION 3 REQUIREMENTS

3.1 Water System

3.1.1 The water system shall be considered as made up of two parts: the utility system and the customer system.

3.1.2 The utility system shall consist of the source facilities and the distribution system and shall include all those facilities of the water system under the complete control of the utility, up to the point where the customer's system begins.

3.1.3 The source shall include all components of the facilities utilized in the production, treatment, storage, and delivery of water to the distribution system.

3.1.4 The distribution system shall include the network of conduits used for the delivery of water from the source to the customer's system.

3.1.5 The customer's system shall include those parts of the facilities beyond the termination of the utility distribution system that are utilized in conveying utility-delivered domestic water to points of use.

3.2 Policy

3.2.1 No water service connection to any premises shall be installed or maintained by the water purveyor unless the water supply is protected as required by state/

provincial laws and regulations and this (*name of legal document*). Service of water to any premises shall be discontinued by the water purveyor if a backflow-prevention assembly required by this (*name of legal document*) is not installed, tested, and maintained, or if it is found that a backflow-prevention assembly has been removed, bypassed, or if an unprotected cross-connection exists on the premises. Service will not be restored until such conditions or defects are corrected.

3.2.2 The customer's system should be open for inspection at all reasonable times to authorized representatives of the (*water or health agency name*) to determine whether cross-connections or other structural or sanitary hazards, including violations of these regulations, exist. When such a condition becomes known, the (*water commissioner or health officer*) shall deny or immediately discontinue service to the premises by providing for a physical break in the service line until the customer has corrected the condition(s) in conformance with state/provincial and city statutes relating to plumbing and water supplies and the regulations adopted pursuant thereto.

3.2.3 An approved backflow-prevention assembly shall be installed on each service line to a customer's water system at or near the property line or immediately inside the building being served; but in all cases, before the first branch line leading off the service line wherever the following conditions exist:

3.2.3a In the case of premises having an auxiliary water supply that is not or may not be of safe bacteriological or chemical quality and that is not acceptable as an additional source by the (*water commissioner or health authority*), the public water system shall be protected against backflow from the premises by installing an approved backflow-prevention assembly in the service line, appropriate to the degree of hazard.

3.2.3b In the case of premises on which any industrial fluids or any other objectionable substances are handled in such a fashion as to create an actual or potential hazard to the public water system, the public system shall be protected against backflow from the premises by installing an approved backflow-prevention assembly in the service line, appropriate to the degree of hazard. This shall include the handling of process waters and waters originating from the utility system that have been subject to deterioration in quality.

3.2.3c In the case of premises having (1) internal cross-connections that cannot be permanently corrected and controlled, or (2) intricate plumbing and piping arrangements or where entry to all portions of the premises is not readily accessible for inspection purposes, making it impracticable or impossible to ascertain whether or not dangerous cross-connections exist, the public water system shall be protected against backflow from the premises by installing an approved backflow-prevention assembly in the service line.

3.2.4. The type of protective assembly required under subsections 3.2.3a, 3.2.3b, and 3.2.3c shall depend upon the degree of hazard that exists as follows:

3.2.4a In the case of any premises where there is an auxiliary water supply as stated in subsection 3.2.3a of this section and it is not subject to any of the following rules, the public water system shall be protected by an approved air-gap separation or an approved reduced-pressure principle backflow-prevention assembly.

3.2.4b In the case of any premises where there is water or substance that would be objectionable but not hazardous to health, if introduced into the public water system, the public water system shall be protected by an approved double check valve assembly.

3.2.4c In the case of any premises where there is any material dangerous to health that is handled in such a fashion as to create an actual or potential hazard to the public water system, the public water system shall be protected by an approved air-gap separation or an approved reduced-pressure principle backflow-prevention assembly. Examples of premises where these conditions will exist include sewage treatment plants, sewage pumping stations, chemical manufacturing plants, hospitals, mortuaries, and plating plants.

3.2.4d In the case of any premises where there are "uncontrolled" cross-connections, either actual or potential, the public water system shall be protected by an approved air-gap separation or an approved reduced-pressure principle backflow-prevention assembly at the service connection.

3.2.4e In the case of any premises where, because of security requirements or other prohibitions or restrictions, it is impossible or impractical to make a complete in-plant cross-connection survey, the public water system shall be protected against backflow from the premises by either an approved air-gap separation or an approved reduced-pressure principle backflow-prevention assembly on each service to the premises.

3.2.4f In the case of any premises where, in the opinion of the (*water commissioner or health officer*), an undue health threat is posed because of the presence of extremely toxic substances, the (*water commissioner or health officer*) may require an air gap at the service connection to protect the public water system. This requirement will be at the discretion of the (*water commissioner or health officer*) and is dependent on the degree of hazard.

3.2.5 Any backflow-prevention assembly required herein shall be a model and size approved by the (*water commissioner or health official*). The term *approved backflow-prevention assembly* shall mean an assembly that has been manufactured in full conformance with the standards established by the American Water Works Association titled:

ANSI/AWWA C510-89—*Standard for Double Check Valve Backflow-Prevention Assembly*, and AWWA C511-89—*Standard for Reduced-Pressure Principle Backflow-Prevention Assembly*, and have met completely the laboratory and field performance specifications of the Foundation for Cross-Connection Control and Hydraulic Research (FCCCHR) of the University of Southern California established by "Specification of Backflow-Prevention Assemblies"—Sec. 10 of the most current issue of the *Manual of Cross-Connection Control*.

Said AWWA and FCCCHR standards and specifications have been adopted by the (*water commissioner or health official*). Final approval shall be evidenced by a "Certificate of Approval" issued by an approved testing laboratory certifying full compliance with said AWWA standards and FCCCHR specifications.

The following testing laboratory has been qualified by the (*water commissioner or health officer*) to test and certify backflow preventers: Foundation for Cross-Connection Control and Hydraulic Research, University of Southern California, University Park, Los Angeles, CA 90089.

Testing laboratories, other than the laboratory listed above, will be added to an approved list as they are qualified by the (*water commissioner or health officer*).

Backflow preventers that may be subjected to backpressure or backsiphonage that have been fully tested and have been granted a certificate of approval by said qualified laboratory and are listed on the laboratory's current list of approved backflow-prevention assemblies may be used without further testing or qualification.

3.2.6 It shall be the duty of the customer–user at any premises where backflow-prevention assemblies are installed to have certified inspections and operational tests made at least once per year. In those instances where the (*water commissioner or health officer*) deems the hazard to be great enough, certified inspections may be required at more frequent intervals. These inspections and tests shall be at the expense of the water user and shall be performed by the assembly manufacturer's representative, (*water department*) personnel, or by a certified tester approved by the (*water commissioner or health officer*). It shall be the duty of the (*water commissioner or health officer*) to see that these tests are made in a timely manner. The customer–user shall notify the (*water commissioner or health officer*) in advance when the tests are to be undertaken so that the customer–user may witness the tests if so desired. These assemblies shall be repaired, overhauled, or replaced at the expense of the customer–user whenever said assemblies are found to be defective. Records of such tests, repairs, and overhaul shall be kept and made available to the (*water commissioner or health officer*).

3.2.7 All presently installed backflow-prevention assemblies that do not meet the requirements of this section but were approved assemblies for the purpose described herein at the time of installation and that have been properly maintained, shall, except for the inspection and maintenance requirements under subsection 3.2.6, be excluded from the requirements of these rules so long as the (*water commissioner or health officer*) is assured that they will satisfactorily protect the utility system. Whenever the existing assembly is moved from the present location, requires more than minimum maintenance, or when the (*water commissioner or health officer*) finds that the maintenance constitutes a hazard to health, the unit shall be replaced by an approved backflow-prevention assembly meeting the requirements of this section.

The foregoing ordinance was first read at the meeting of the (*name of governing body*) of the (*city, county, state, or province*) of _____ on the _____ day of _____, 20_____ and adopted by the following called vote on motion of (*official*).

Ayes: _____
No's: _____
Abstaining: _____
Absent: _____

Approver _____
(Official Title)

ATTEST:

_____
(Clerk or Secretary)

(Seal)

This page intentionally blank.

Appendix C

Common Symbols

Air-Gap Fitting	Gate Valve
Reduced-Pressure Principle Backflow-Prevention Assembly RPBA or	Air Gap
Atmospheric Vacuum Breaker ABV	Pressure-Reducing Valve
Check Valve	Pressure Relief Valve
Double Check Valve Assembly DCV	Pressure Vacuum Breaker PVB
Ejector or Aspirator Unit	Irrigation System

This page intentionally blank.

AWWA MANUAL M14

Appendix D

AWWA Policy Statement on Cross-Connections

Adopted by the Board of Directors Jan. 26, 1970, revised June 24, 1979, reaffirmed June 10, 1984, and revised Jan. 28, 1990, and Jan. 21, 2001.

The American Water Works Association (AWWA) recognizes water purveyors have the responsibility to supply potable water to their customers. In the exercise of this responsibility, water purveyors or other responsible authorities must implement, administer, and maintain ongoing backflow prevention and cross-connection control programs to protect public water systems from the hazards originating on the premises of their customers and from temporary connections that may impair or alter the water in the public water systems. The return of any water to the public water system after the water has been used for any purpose on the customer's premises or within the customer's piping system is unacceptable and opposed by AWWA.

The water purveyor shall assure that effective backflow-prevention measures, commensurate with the degree of hazard, are implemented to ensure continual protection of the water in the public water distribution system. Customers, together with other authorities, are responsible for preventing contamination of the private plumbing system under their control and the associated protection of the public water system.

If appropriate backflow-prevention measures have not been taken, the water purveyor shall take or cause to be taken necessary measures to ensure that the public water distribution system is protected from any actual or potential backflow hazard. Such action would include the testing, installation, and continual assurance of proper operation and installation of backflow-prevention assemblies, devices, and methods commensurate with the degree of hazard at the service connection or at the point of cross-connection or both. If these actions are not taken, water service shall ultimately be eliminated.

To reduce the risk private plumbing systems pose to the public water distribution system, the water purveyor's backflow-prevention program should include public education regarding the hazards backflow presents to the safety of drinking water and should include coordination with the cross-connection efforts of local authorities, particularly health and plumbing officials. In areas lacking a health or plumbing enforcement agency, the water purveyor should additionally promote the health and safety of private plumbing systems to protect its customers from the hazards of backflow.

AWWA MANUAL M14

Appendix E

Abbreviations of Agencies and Organizations

NOTE: Abbreviations used refer to reference materials issued by the organization identified below:

ABPA American Backflow Prevention Association
3829 Old College Rd., Bryan, TX 77801-4112,
(979) 846-7606

AHAM Association of Home Appliance Manufacturers
1111 19th St., N.W., Suite 402, Washington, DC 20036,
(202) 872-5955

ANSI American National Standards Institute
25 W. 43rd St., Fourth Floor, New York, NY 10036-7406,
(212) 642-4900

ASME American Society of Mechanical Engineers
Three Park Ave., New York, NY 10016,
(212) 591-7740

ASSE American Society of Sanitary Engineering
901 Canterbury Rd., Suite A, Westlake, OH 44145-7201,
(440) 835-3040

ASTM American Society for Testing and Materials
100 Barr Harbor Dr., West Conshohocken, PA 19428,
(610) 832-9585

AWWA	American Water Works Association 6666 West Quincy Ave., Denver, CO 80235, (303) 794-7711
CISPI	Cast Iron Soil Pipe Institute 5959 Shallowford Rd., Suite 419, Chattanooga, TN 37421, (423) 892-0137
CSA	Canadian Standards Association International Etobicoke (Toronto) 178 Rexdale Blvd., Etobicoke, ON, M9W 1R3, Canada, (800) 463-6727
CS&PS	Commercial Standards and Product Standards Superintendent of Documents, US Govt. Printing Office, 732 N. Capitol St., N.W., Washington, DC 20401
FM	Factory Mutual Global PO Box 7500, Johnston, RI 02919, (401) 275-3000
FS	Federal Specifications Superintendent of Documents, US Govt. Printing Office, 732 N. Capitol St., N.W., Washington, DC 20402
IAPMO	International Association of Plumbing and Mechanical Officials (Uniform Plumbing Codes) (UPC) 5001 E. Philadelphia St., Ontario, CA 91761-2816, (909) 472-4100
NFPA	National Fire Protection Association 1 Batterymarch Park, Quincy, MA 02169-7471, (800) 344-3555
NSF	NSF International 789 N. Dixboro Rd., PO Box 130140, Ann Arbor, MI 48113-0140, (800) NSF-MARK
PDI	Plumbing and Drainage Institute 800 Turnpike St., Suite 300, North Andover, MA 01845, 1 (978) 557-0720, 1 (800) 589-8956
USC	University of Southern California Foundation for Cross-Connection Control and Hydraulic Research KAP-200 University Park MC-2531, Los Angeles, CA 90089-2531, (213) 740-2032
UL	Underwriters' Laboratories, Inc. 333 Pfingsten Rd., Northbrook, IL 60062, (708) 272-8800

AWWA MANUAL M14

Appendix F

Web Sites for Backflow-Prevention Incidents

Additional reports of backflow incidents are available on the following Web sites or through many state departments of environmental protection agencies.

www.epa.gov/ogwdw/tcr/pdf/ccrwhite.pdf
www.epa.gov/safewater/crossconnection.html
www.abpa.org
www.classicbackflow.com
www.treeo.ufl.edu/backflow/casehist.asp
www.usc.edu/dept/fccchr/ccvlib

This page intentionally blank.

Glossary

Absolute pressure Gauge pressure plus atmospheric pressure. It is measured in units of pounds per square inch absolute (psia).

Administrative authority The official office, board, department or agency authorized by law to administer and enforce regulation and or code requirements. This includes a duly authorized representative of the administrative authority.

Air gap (AG) The unobstructed vertical distance through free atmosphere between the lowest effective opening from any pipe or faucet conveying water or waste to a tank, plumbing fixture, receptor, or other assembly and the flood level rim of the receptacle. These vertical, physical separations must be at least twice the effective opening of the water supply outlet, never less than 1 in. (25 mm) above the receiving vessel flood rim. Local codes, regulations, and special conditions may require more stringent requirements.

Air-gap fitting A physical device engineered to produce a proper air-gap separation as defined above.

Approved Accepted by the authority having jurisdiction as meeting an applicable standard, specification, requirement, or as suitable for the proposed use.

Assembly An assemblance of one or more approved body components and including approved shutoff valves.

Atmospheric pressure The pressure exerted by the atmosphere at any point. Such pressure decreases as the elevation of the point above sea level increases. One atmosphere is equivalent to 14.7 psi (101.4 kPa), 29.92 in. (760 mm) of mercury, or 33.9 ft (10.1 m) of water column at average sea level.

Atmospheric vacuum breaker (AVB) The AVB consists of a float check, a check seat, and an air-inlet port. A shutoff valve immediately upstream may or may not be an integral part of the device. The AVB is designed to allow air to enter the downstream water line to prevent backsiphonage. This unit may never be subjected to a backpressure condition or have a downstream shutoff valve, or be installed where it will be in continuous operation for more than 12 hours.

Authority having jurisdiction The agency, organization, office, or individual responsible for approving materials, equipment, work, installation, or procedure.

Auxiliary water supply Any water supply on or available to the premises other than the water purveyor's approved public water supply. These auxiliary waters may include water from another water purveyor's public potable water supply or any natural source(s), such as a well, lake, spring, river, stream, harbor, and so forth; or used waters, reclaimed waters, recycled waters, or industrial fluids. These waters may be contaminated or polluted or they may be objectionable and constitute an unacceptable water source over which the water purveyor does not have sanitary control.

Backflow The undesirable reversal of flow of a liquid, gas, or other substance in a potable water distribution piping system as a result of a cross-connection.

Backflow preventer An assembly, device, or method that prohibits the backflow of water into potable water supply systems.

Backpressure A pressure, higher than the supply pressure, caused by a pump, elevated tank, boiler, air/steam pressure, or any other means, which may cause backflow.

Backsiphonage A type of backflow where the upstream pressure to a piping system is reduced to a subatmospheric pressure.

Ball valve A valve with a spherical gate providing a tight shutoff. Ball valves on backflow assemblies shall be fully ported and resilient seated.

Barometric loop A looped piping arrangement 35 ft (11 m) in height in which the water flow goes over the loop at the top. This method of backflow prevention is only capable of protecting against backsiphonage, since backpressure could drive water backward over the loop.

Certified backflow-prevention assembly tester A person who has demonstrated competence to test, repair, and maintain backflow-prevention assemblies as evidenced by certification that is recognized by the approving authority.

Consumer The owner, operator, or customer having a service from a public potable water system.

Critical level A reference line representing the level of the check valve seat within a backsiphonage control unit. It is used to establish the height of the unit above the highest outlet or flood level rim. If it is not marked on the backflow preventer, the bottom of the assembly is the critical level.

Cross-connection A connection or a potential connection between any part of a potable water system and any other environment containing other substances in a manner that, under any circumstances, would allow such substances to enter the potable water system. Other substances may be gases, liquids, or solids, such as chemicals, water products, steam, water from other sources (potable or nonpotable), or any matter that may change the color or add odor to the water. Bypass arrangements, jumper connections, removable sections, swivel or changeover assemblies, or any other temporary or permanent connecting arrangement through which backflow may occur are considered to be cross-connections.

Cross-connection control A program to eliminate, monitor, protect, and prevent cross-connections from allowing backflow.

Direct cross-connection A cross-connection that is subject to both backsiphonage and backpressure.

Disk The part of a valve that actually closes off flow.

Double check detector backflow-prevention assembly (DCDA) A specially designed backflow assembly composed of a line-size-approved double check valve assembly with a bypass containing a specific water meter and an approved double check valve assembly. The meter shall register accurately for only very low rates of flow up to 3 gpm and shall show a registration for all rates of flow. This assembly shall only be used to protect against a non-health hazard (i.e., a pollutant).

Double check valve assembly (DC or DCVA) A complete assembly consisting of two internally loaded, independently operating check valves, located between two tightly closing resilient-seated shutoff valves with four properly placed resilient-seated test cocks. This assembly shall only be used to protect against a non-health hazard (i.e., a pollutant).

Effective opening The minimum cross-sectional area at the point of water supply discharge, measured or expressed in terms of the diameter of a circle, or if the opening is not circular, the diameter of a circle of equivalent cross-sectional area.

Electrolysis The corrosion resulting from the flow of electric current.

Expansion tank A tank used for safely controlling the expansion of water.

Field testing A procedure to determine the operational and functioning status of a backflow preventer.

Fire department connection (FDC or Siamese connection) A connection through which a fire department can introduce supplemental water with or without the addition of other chemical fire-retarding agents by the means of a pump into a sprinkler system, standpipe, or other fire-suppression system.

Fire protection systems (water based)

Antifreeze system: A wet-pipe sprinkler system containing antifreeze.

Combined dry pipe-preaction system: A sprinkler system containing air under pressure with a supplemental detection system installed in the area of the sprinklers. The detection system actuates tripping devices that open water inlet and air exhaust valves, which generally precedes the opening of the sprinklers. The detection system additionally serves as a fire alarm system.

Deluge system: A sprinkler system having open sprinkler heads connected to a water supply. The sprinkler system piping is dry until the fire-detection system opens the water supply valve to the system.

Dry-pipe system: A sprinkler system containing air or nitrogen under pressure and connected to a water supply. A sprinkler head opening allows the air or nitrogen to be released from the system and water to enter the system. Dry-pipe systems are to be maintained dry at all times. *Exception: During nonfreezing conditions, the system can be left wet if the only other option is to remove the system from service while waiting for parts or during repair activities.*

Foam water sprinkler and spray systems: A special fire protection system pipe connected to a source of foam concentrate and to a water supply. The system may discharge the foam agent before, after, or with the water over the area to be protected.

Preaction system: A sprinkler system containing air that may or may not be under pressure and connected to a water supply but having a supplemental detection system in the area of the sprinklers that would open a supply valve, allowing water to flow into the system and to be discharged by any open sprinkler.

Sprinkler system: A system of underground and overhead piping hydraulically designed and constructed to which sprinkler heads are attached for extinguishing fire.

Standpipe system: A piping system having valves, hose connections, and allied equipment installed within a premises, building, or structure where the hose connections are located in a manner to discharge water through an attached hose and nozzle to extinguish a fire. These systems may be wet or dry and may or may not be directly connected to a drinking water supply system. They may also be combined with a sprinkler system. There are three classes of standpipe systems. Class I service provides 2½-in. hose stations from a standpipe or combined riser. Class II service provides 1½-in. hose stations from a standpipe, combined riser, or sprinkler system. *Exception: A minimum 1-in. hose may be used for Class II Light Hazard Occupancies if investigated, listed, and authorized by the authority having jurisdiction.* Class III service provides 1½-in.

and 2½-in. hose connections or 1½-in. or 2½-in. hose stations from a standpipe or combination riser.

Wet-pipe system: A sprinkler system containing water and connected to a water supply.

NOTE: For further information on fire protection systems, refer to National Fire Protection Association publications.

Flood level rim That level from which liquid in plumbing fixtures, appliances, or vats could overflow to the floor, when all drain and overflow openings built into the equipment are obstructed.

Gauge pressure The pressure at a point of a substance (gas or liquid) above that of the atmosphere.

Health hazard (high hazard) A cross-connection or potential cross-connection involving any substance that could, if introduced into the potable water supply, cause death or illness, spread disease, or have a high probability of causing such effects.

Indirect cross-connection A cross-connection that is subjected to backsiphonage only.

Inspection A visual examination of backflow-protection equipment, materials, workmanship or portion thereof to verify installation and operational performance.

Inspector An individual working for or under the authority having jurisdiction empowered to ensure code compliance.

Internal protection Fixture isolation and/or isolation of an area or zone. Protection at the fixture means installing an approved backflow preventer at the source of the potential hazard within a specific area.

Irrigation water Water utilized for plant life.

Listed-classified-approved Materials, equipment, fixtures, and other products included in a list published by an agency or organization that has successfully evaluated the item and determined compliance with the agency's established material and/or performance standards.

Maintenance Work performed or repairs made to keep equipment operable and in compliance.

Needle valve A valve that has a small opening that is closed or opened by a needle-like spindle. Used for fine control.

Non-health hazard (low hazard) A cross-connection or potential cross-connection involving any substance that generally would not be a health hazard but would constitute a nuisance or be aesthetically objectionable if introduced into the potable water supply.

Plumbing system All potable water and distribution pipes, fixtures, traps, drainage pipe, gas pipe, water treating or using equipment, vent pipe, including joints, connections, devices, receptacles, and appurtenances within the property lines of a premises.

Pollution (*see* Non-health hazard)

Potable water Water that is safe for human consumption as described by the public health authority having jurisdiction.

Premises isolation Preventing backflow into a public water system from a user's premises by installing a suitable backflow preventer at all the user's potable water connections (*see* Service protection).

Pressure vacuum-breaker assembly (PVB) An assembly consisting of an independently operating, internally loaded check valve, an independently operating, loaded air-inlet valve located on the discharge side of the check valve, with properly located resilient-seated test cocks and tightly closing resilient-seated shutoff valves attached at each end of the assembly designed to be operated under pressure for prolonged periods of time to prevent backsiphonage. The pressure vacuum breaker may not be subjected to any backpressure.

Reclaimed water Water that, as a result of treatment of wastewater, is suitable for a direct beneficial use or a controlled use that would not otherwise occur and is not safe for human consumption.

Reduced-pressure principle backflow-prevention assembly (RP or RPBA or RPA or RPZ) A complete assembly consisting of a mechanical, independently acting, hydraulically dependent relief valve, located between two independently operating, internally loaded check valves that are located between two tightly closing resilient-seated shutoff valves with four properly placed resilient-seated test cocks.

Reduced-pressure principle detector backflow-prevention assembly (RPDA) A specially designed backflow assembly composed of a line-size approved reduced-pressure principle backflow-prevention assembly with a bypass containing a specific water meter and an approved reduced-pressure principle backflow-prevention assembly. The meter shall register accurately for only very low rates of flow up to 3 gpm and shall show a registration for all rates of flow. This assembly shall be used to protect against a non-health hazard (i.e., a pollutant) or a health hazard (i.e., a contaminant). The RPDA is primarily used on fire sprinkler systems.

Service connection A piping connection between the water purveyor's main and a user's system.

Service protection Containment protection or secondary protection refers to the backflow protection installed on the water supply line to a premises as close to the service connection to the public water system as possible (*see* Premises isolation).

Sewage Liquid waste containing human, animal, chemical, or vegetable matter in suspension or solution.

Single check valve A single check valve is a directional flow control valve, but not an approved backflow preventer.

Spill-resistant pressure vacuum-breaker backsiphonage-prevention assembly (SVB) A backflow assembly containing an independently operating, internally loaded check valve and independently operating, loaded air-inlet valve located on the discharge side of the check valve. The assembly is to be equipped with a properly located resilient-seated test cock, a properly located bleed/vent valve, and tightly closing resilient-seated shutoff valves attached at each end of the assembly. This assembly is designed to protect against a non-health hazard (i.e., a pollutant) or a health hazard (i.e., a contaminant) under backsiphonage condition only.

Static water level The height measurement of a liquid at rest within a vessel.

Submerged inlet An inlet pipe opening that is below the flood level rim of the receptacle.

Test equipment An electronic or mechanical instrument recognized by the authority having jurisdiction to field-test the operational performance of a backflow preventer (*see* Field testing).

Union A three-part coupling device used to join pipe.

Valve seat The port(s) against or into which a disk or tapered stem is pressed or inserted into to shut down flow.

Velocity The speed of motion in a given direction.

Water purveyor The owner or operator of a public (or private) potable waterworks systems.

Index

NOTE: *f.* indicates figure; *t.* indicates table.

AG. *See* Air gap
Air gap, 42, 42*f.*
   applications, 42–44, 43*f.*, 44*f.*
   for auxiliary water systems, 63–65
   for beverage-bottling plants and breweries, 65
   for chemical plants and facilities, 66–67
   for dairies and cold-storage plants, 67
   in distribution systems, 80–81
   and dry-pipe nonpressurized fire-suppression systems, 68
   for dye plants, 68
   for film laboratories, 68–69
   for fire hydrants, 69
   for food processing facilities, 65–66
   for irrigation systems, 70–71
   for laundries and dye works, 71–72
   for medical facilities and mortuaries, 70
   for metal facilities, 72–73
   for multistoried buildings, 73
   for oil and gas properties, 74
   for paper and paper-product plants, 75
   for plating plants and facilities, 75–76
   for radioactive materials facilities, 76
   for reclaimed or recycled water, 76–77
   for residential water services, 77
   for restricted facilities, 78
   for steam boiler plants, 79
   in water treatment plants, 83
   for water-hauling equipment, 79
   for wet-pipe fire sprinkler systems, 79–80
Area isolation, 83–84, 84*f.*
Atmospheric pressure, 29
Atmospheric vacuum breakers, 56, 57*f.*
Auxiliary water systems, 63–64
AVB. *See* Atmospheric vacuum breakers

Backflow
   risk assessment, 37–38
   types of, 31–34
   See also Backpressure, Backsiphonage
Backflow devices, 56
   atmospheric vacuum breaker, 56, 57*f.*
   contrasted with backflow-prevention assemblies, 41–42
   dual check, 57–58, 58*f.*
   dual check with atmospheric vent, 58–59
   See also Backflow-prevention assemblies
Backflow incident response plans, 25–27
Backflow preventers, 41–42
   air gap, 42–44, 42*f.*, 43*f.*, 44*f.*, 63–80, 83
   assemblies vs. devices, 41–42
   factors in selecting, 41
   See also Backflow devices, Backflow-prevention assemblies
Backflow-prevention assemblies
   assessing effectiveness of, 38–40, 40*t.*
   contrasted with backflow devices, 41–42
   double check detector assembly, 52, 79–80
   double check valve assembly, 49–52, 49*f.*, 50*f.*, 51*f.*, 63–65, 68–70, 72–73, 77–80, 83
   field-testing equipment, 60–61
   field-testing procedure, 59–60
   inventory of, 22–23
   maintenance, 61
   ownership of, 14–15
   pressure vacuum-breaker assembly, 52–55, 53*f.*, 54*f.*, 71
   quality assurance, 20
   reduced-pressure principle assembly, 44–48, 45*f.*, 46*f.*, 47*f.*, 48*f.*, 65–80, 83
   reduced-pressure principle detector assembly, 48–49, 68, 79–80
   spill-resistant vacuum breakers, 55–56
   standards, 19–20
   test reports, 23
   tester training and certification, 17–19
   tester's responsibility, 61
   testing, 15, 17, 59–61
   testing and continuous water service, 59
   testing frequency, 59
   types of, 40*t.*
   in water treatment plants, 82–83
   See also Backflow devices
Backpressure, 32–34
   caused by elevated piping, 34
   caused by pressurized containers, 34, 35*f.*
   caused by pumps, 33–34, 36*f.*
   caused by thermal expansion, 34
Backsiphonage, 31
   caused by high rate of water withdrawal, 31, 32*f.*

caused by reduced pressure on suction side of booster pump, 31, 33f.
caused by shutdown of water system, 31, 33f.
Bacteria, 5
Barometric loop, 30
Beverage-bottling plants and breweries, 65

Canadian regulatory agencies, 1
Chemical plants and facilities, 66–67
Chemicals, 5–6
Cold-storage plants, 67
Consumer Confidence Reports, 24
Cross-connection control programs
  authority and administrative responsibility, 13
  backflow incident reports, 23
  backflow incident response plans, 25–27
  backflow-prevention assembly testing and reports, 15, 17, 23
  budgets and funding, 27
  common law doctrines, 8–9
  and Consumer Confidence Reports, 24
  contractual responsibilities, 6–7
  coordination with local authorities, 20–22
  correspondence retention, 23
  database management, 23
  enforcement actions, 15–16
  health aspects, 3–6
  internal fixture protection, 14
  inventory of backflow-prevention assemblies, 22–23
  joint programs, 21–22
  mandatory service protection, 13–14
  ownership of service protection backflow-prevention assemblies, 14–15
  public education, 23–25
  quality assurance, 20
  record keeping, 22–23
  and regulations, 1–3, 7–8
  responsible parties (other than water purveyors), 11–12
  safety considerations, 25
  service policy, 13–14, 20–21
  spreadsheets, 23
  standards for backflow-prevention assemblies and field-test equipment, 19–20
  training and certification, 17–19, 20
  water purveyors' general responsibilities, 11–12
  water purveyors' risk assessment, 16–17, 22
Cross-connections
  defined, 1
  risk assessment, 35

Dairies, 67
DC. See Double check valve backflow-prevention assemblies
DCDA. See Double check detector backflow-prevention assemblies
Distribution systems, 80–81
Dockside watering points, 72
Double check detector backflow-prevention assemblies, 52
  for wet-pipe fire sprinkler systems, 79–80
Double check valve backflow-prevention assemblies, 49, 49f.
  applications, 50, 51f.
  for auxiliary water systems, 63–65
  for beverage-bottling plants and breweries, 65
  for dry-pipe pressurized and preaction fire-suppression systems, 68
  for fire sprinkler systems, 69–70
  function, 49–50, 50f.
  installation criteria, 51–52
  for metal facilities, 72–73
  for multistoried buildings, 73
  for residential water services, 77
  for solar domestic hot-water systems, 78–79
  in water treatment plants, 83
  for wet-pipe fire sprinkler systems, 79–80
Dry-pipe nonpressurized fire-suppression systems, 68
Dry-pipe pressurized and preaction fire-suppression systems, 68
Dual check, 57–58, 58f.
Dual check with atmospheric vent, 58–59
Dye plants, 68
Dye works, 71–72

Film laboratories, 68–69
Fire hydrants, 69
Fire sprinkler systems, 69–70
  single-family, 77–78
  wet-pipe, 79–80
Fire-suppression systems, 68, 74–75
Food processing facilities, 65–66

Giardia, 5

Hydraulic grade line, 30–31

Internal fixture protection, 14
Irrigation systems, 70–71

Laundries, 71–72

Mandatory service protection, 13–14
Marine facilities, 72

Medical facilities, 70
Metal facilities, 72–73
Mortuaries, 70
Multiple-barrier approach, 3–4
Multistoried buildings, 73

National Primary Drinking Water
 Regulations, 8
NPDWRs. *See* National Primary Drinking
 Water Regulations

Oil and gas properties, 74

Paper and paper-product plants, 75
Plating plants and facilities, 75–76
Plumbing codes, 4
Potable water, 1, 2
 defined, 4
Pressure vacuum-breaker assemblies,
 52, 53f.
 applications, 54, 54f.
 function, 52–53, 53f.
 installation criteria, 54–55
 for irrigation systems, 70–71
Public water systems, 2
PVB. *See* Pressure vacuum-breaker
 assemblies

Radioactive materials facilities, 76
Reclaimed or recycled water, 76–77.
 See also Used water
Reduced-pressure principle
 backflow-prevention assemblies, 44–45,
 45f.
 applications, 47, 48f.
 for beverage-bottling plants and
  breweries, 65
 for chemical plants and facilities, 66–67
 for dairies and cold-storage plants, 67
 and dry-pipe nonpressurized
  fire-suppression systems, 68
 for dry-pipe pressurized and preaction
  fire-suppression systems, 68
 for dye plants, 68
 for film laboratories, 68–69
 for fire hydrants, 69
 for fire sprinkler systems, 69–70
 for food processing facilities, 65–66
 function, 45–47, 46f., 47f.
 installation criteria, 47–48
 for irrigation systems, 70–71
 for laundries and dye works, 71–72
 for marine facilities and dockside
  watering points, 72
 for medical facilities and mortuaries, 70

 for metal facilities, 72–73
 for multistoried buildings, 73
 for oil and gas properties, 74
 for paper and paper-product plants, 75
 for plating plants and facilities, 75–76
 for radioactive materials facilities, 76
 for reclaimed or recycled water, 76–77
 for residential water services, 77
 for restricted facilities, 78
 for solar domestic hot-water
  systems, 78–79
 for steam boiler plants, 79
 for water-hauling equipment, 79
 for wet-pipe fire sprinkler systems, 79–80
Reduced-pressure principle detector
 backflow-prevention assemblies, 48–49
 in distribution systems, 80–81
 and dry-pipe nonpressurized
  fire-suppression systems, 68
 in water treatment plants, 83
 for wet-pipe fire sprinkler systems, 79–80
Regulations, 7–8
 responsibility for, 1–3
Residential water services, 77
Restricted facilities, 78
Risk assessment, 16–17, 34–35, 38
 of actual connections, 37
 of backflow conditions, 37–38
 of cross-connections, 35
 of potential connections, 37
 records, 22
 and used water, 38
RP. *See* Reduced-pressure principle
 backflow-prevention assemblies
RPDA. *See* Reduced-pressure principle
 detector backflow-prevention assemblies

Safe Drinking Water Act, 1, 2, 8
 reporting requirements, 8
SDWA. *See* Safe Drinking Water Act
Service containment, 83–84, 84f.
Solar domestic hot-water systems, 78–79
Spill-resistant vacuum breakers, 55–56
Steam boiler plants, 79

Treatment plants, 81–82
 cross-connection control, 83f.
 recommended protection at fixtures and
  equipment, 82–83, 82t.
 service containment and area isolation,
  83–84, 84f.

Used water, 38, 64. *See also* Reclaimed or
 recycled water

Vacuum
  defined, 29–30
  schematic, 30f.
Venturi effect, 30

Water pressure, 29
Water purveyors
  cross-connection control in own offices and work areas, 84
  general cross-connection control responsibilities, 11–12
  risk assessment, 16–17, 22

Water quality
  and chemicals, 5–6
  common law doctrines, 8–9
  contractual responsibilities, 6–7
  health aspects, 3–6
  legal aspects, 6–9
  microbiological considerations, 5
  and physical hazards, 6
  and regulations, 1–3, 7–8
Water-hauling equipment, 79
Waterborne diseases, 5
Wet-pipe fire sprinkler systems, 79–80